Elektrische Energiespeichersysteme
Herausgeber: Prof. Dr.-Ing. Kai Peter Birke

Band 6

Optimal Energy Management Strategies for Reconfigurable Batteries

Von der Fakultät Informatik, Elektrotechnik und Informationstechnik der Universität Stuttgart zur Erlangung der Würde eines Doktor-Ingenieurs (Dr.-Ing.) genehmigte Abhandlung

Vorgelegt von

Nejmeddine Bouchhima

geboren in Sfax

Hauptberichter:	Prof. Dr.-Ing. Kai Peter Birke
Mitberichter:	Prof. Dr.-Ing. Julia Kowal

Tag der mündlichen Prüfung: 10. Juli 2020

Institut für Photovoltaik der Universität Stuttgart

2021

Bibliografische Information der Deutschen Nationalbibliothek
Die Deutsche Nationalbibliothek verzeichnet diese Publikation in der Deutschen Nationalbibliografie; detaillierte bibliografische Daten sind im Internet über http://dnb.d-nb.de abrufbar.

1. Aufl. - Göttingen: Cuvillier, 2021

Zugl.: Stuttgart, Univ., Diss., 2020

Nonnenstieg 8, 37075 Göttingen
Telefon: 0551-54724-0
Telefax: 0551-54724-21
www.cuvillier.de

1. Auflage, 2021
Gedruckt auf umweltfreundlichem, säurefreiem Papier aus nachhaltiger Forstwirtschaft

ISBN 978-3-7369-7414-2
eISBN 978-3-7369-6414-3

Contents

Abstract

Multicell Lithium-ion batteries are widely used in electric vehicles and stationary storage systems due to their high energy and power density. However, due to the capacity and internal resistance disparity, the battery cells need to be balanced. The reconfigurable battery architecture is considered a promising solution to overcome the cell imbalance, as the architecture enables the activation and deactivation of each battery cell based on its current state. The scope of this work is to develop energy management strategies for the battery architecture mentioned above in order to maximize the energy efficiency of the battery. Two mathematical models are introduced in order to describe the dependency of the cell depth of discharge and the battery balancing state on the number of active cells. These models are used to formulate the dynamic cell switching process as a nonlinear dynamic optimization problem. To calculate the global optimum, an optimization algorithm based on the forward dynamic programming technique is introduced. This strategy is a noncausal strategy as it relies on the availability of battery load information over the optimization horizon, e.g., a priori known driving cycle. Thereafter, the dynamic optimization problem is transformed into a static optimization problem in order to develop causal strategies. The latter are only able to achieve a local optimum because they do not exploit information about the entire driving cycle. In a second step, simulation studies are conducted to assess the performance of these strategies using an equivalent electrical battery model. Results reveal that using the optimization-based strategies presented in this work can improve the energy efficiency of the battery compared to batteries with a static cell connection and passive balancing. In contrast to the conventional rule-based strategy for reconfigurable batteries, the developed strategies also show robustness to capacity variation inside the battery. Whereby, the efficiency enhancement achieved by the causal strategies is only slightly worse than the results of the noncausal strategy which performs poorly in terms of computational effort and real-time capability. Therefore, the causal strategies, which show a better trade-off between computational effort and accuracy than noncausal strategy, are better suitable for real-time applications.

Zusammenfassung

Lithium-Ionen-Batterien werden aufgrund ihrer hohen Energie- und Leistungsdichte häufig in Elektrofahrzeugen und stationären Speichersystemen eingesetzt. Allerdings müssen die Batteriezellen aufgrund der Kapazitätsstreuung ausgeglichen werden. Die Einzelzellschalter Topologie, die die Aktivierung und Deaktivierung jeder Batteriezelle basierend auf ihrem aktuellen Zustand ermöglicht, gilt als vielversprechende Lösung, um die Inhomogenität der Batteriezellen zu überwinden. Ziel dieser Arbeit ist die Entwicklung von optimalen Energiemanagement-Strategien für die Einzelzellschalter Batterie, um die Energieeffizienz des Systems zu maximieren. Zuerst werden zwei mathematische Modelle erstellt, die die Abhängigkeit des Entladezustands jeder Zelle und des Ausgleichszustands der Batterie von der Anzahl der aktiven Zellen beschreiben. Diese Modelle werden verwendet, um die dynamische Verschaltung der Zellen als nichtlineares dynamisches Optimierungsproblem zu formulieren. Für die Berechnung des globalen Optimums wird ein Optimierungsalgorithmus, basierend auf der dynamischen Programmierung Methode, entwickelt. Diese Strategie ist eine nicht-kausale Strategie, da die Batterieleistungsanforderungen über den Optimierungshorizont für die Strategie bekannt sein müssen. In einer Simulationsumgebung werden diese Informationen eines vorgegebenen Fahrprofils entnommen.

Als nächster Schritt werden zwei statische Optimierungsprobleme von dem dynamischen Optimierungsproblem abgeleitet, damit die Entwicklung von kausalen Energiemanagement-Strategien ermöglicht werden kann. Diese Strategien berechnen ein lokales Optimum, da sie keine Informationen über den gesamten Fahrzyklus verwenden.

Die Validierung der entwickelten Strategien wird anhand von Entladeszenario Simulationen durchgeführt. Für diese Simulationen werden realistische Fahrzyklen und ein Ersatzschaltbild Batteriemodel mit normalverteilten Zellkapazitäten verwendet. Der Fokus liegt bei der Auswertung der Ergebnisse auf der Güte der Strategien hinsichtlich Steigerung des Batteriewirkungsgrades. Die Ergebnisse zeigen, dass die entwickelten optimierungsbasierten Strategien den Energiewirkungsgrad der Batterie im Vergleich zu herkömmlichen Batterien mit passivem Balancing deutlich verbessern. Außerdem zeigen die entwickelten Strategien im Gegensatz zur herkömmlichen regelbasierten Strategie für Einzelzellschalter Batterien eine Robustheit gegenüber Verschlechterung der Kapazitätsstreuung innerhalb der Batterie. Wobei ist die Wirkungsgradsteigerung durch die kausalen Strategien nur geringfügig schlechter als die Ergebnisse der nicht-kausalen Strategie, die in Bezug auf Echtzeitfähigkeit einen hohen Rechenaufwand aufweist. Deswegen sind die kausalen Strategien, die einen besseren Kompromiss zwischen Rechenaufwand und Genauigkeit als die nicht-kausale Strategie zeigen, besser für Echtzeitanwendungen geeignet.

Chapter 1

Introduction

1.1 Motivation

Battery electrical vehicles (BEVs) are widely regarded as a promising solution for the future as they are a great way to improve mobility and reduce pollution. The increased awareness of the environmental impact of conventional internal combustion engine (ICE) vehicles and climate change consequences makes BEVs increasingly more attractive and popular. The main power sources of these vehicles are energy storage systems made up of Lithium-ion technology. The chief features that make this technology attractive nowadays are its high voltage, high energy density, high efficiency, and long cycle. To satisfy the system requirements on operating voltage, power, and driving range, the battery consists of a large number of single cells, which are connected in series and parallel. While connecting cells in parallel leads to an increase in capacity, the battery output voltage increases when cells are connected in series. Due to its chemistry, a Lithium-ion cell has a defined operating window. If it is operated outside this window, it will be permanently damaged or destroyed. For this reason, all cells are monitored by a battery management system that ensures a safe operation. Unfortunately, the battery cells are not identical in terms of electric characteristics such as capacity and internal resistance. This cell-to-cell variation arises due to manufacturing process tolerances. Despite the end of the production classification of battery cells, the capacity spread remains by 1% to 3% at the beginning of life [1]. In addition, the cell disparity is exacerbated over battery life due to operation conditions, among which temperature gradients [2]. Since the useful capacity is limited by the lowest capacity cell, its efficiency depends on the capacity spread. Figure 1.1 shows how the cell with the lowest capacity limits the electrical energy extracted from the storage system.

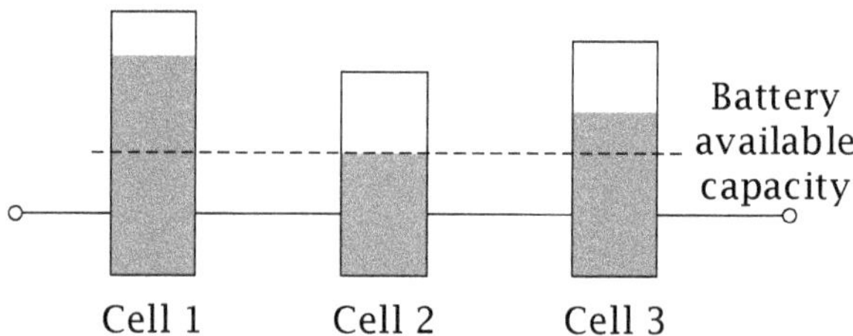

Figure 1.1: Schematic representation of interconnected battery cells with different capacities and states of charge. The available battery capacity is limited by the weakest cell.

Due to the limited operating range of the cells, cell 2 with the lowest charge limits the entire battery pack. If this cell is completely discharged, no further energy shall be taken from the battery pack, as otherwise deep discharging and thus damage to this cell will occur. The remaining load in the other cells remains unused. Accordingly, the cell with the largest state of charge limits the battery pack during the charging process. To increase BEV competitiveness, the current research focuses on improving the efficiency of electric driving and energy storage. To this end, various approaches for equalizing imbalance battery cells use either resistors to dissipate excess load in the form of heat or components to transfer energy from one cell to the other. The amount of energy shift between cells is limited through the used electronic components, such as capacitors and transformers [3].

In this work, a new approach to balance cells is examined. This approach is based on a modular battery system consisting of cell modules connected in series, denoted reconfigurable battery. Thanks to semiconductor switches, these modules can be connected and disconnected during the operation in order to equalize the state of charge of the cells. For example, cell 2, shown in figure 1.1, is excluded from the discharging process once it is bypassed. As a result, the states of charge of the other cells are approaching the charge level of cell 2. The new degree of freedom available in reconfigurable battery, namely the terminal voltage of the battery that can be controlled by bypassing battery cells, should be exploited to achieve the best possible system efficiency. This is even more important because the performance of reconfigurable batteries in terms of efficiency and lifetime not only depends on the cell interconnection topology but rather on the energy management strategy, which controls the cell state and decides when and which cell should be bypassed. Despite this, these reconfigurable battery systems are widely discussed in the literature from a hardware realization point of view. Energy management strategies for reconfigurable batteries are often developed using rule-based controls [4]. On the other hand, a lot of research has been carried out in the field of energy management strategies for electric hybrid vehicles in the last few years. These strategies are typically divided into rule-based strategies and into optimization-based controls, including instantaneous optimization and dynamic programming optimization [5] and [6]. Rule-based controls are widely applied in real-time applications, particularly due to their low computation and memory resources required. Authors in [7] and [8] presented rule-based controls using fuzzy logic. As the rules were not model-based developed, the control strategies are thus not scalable. Furthermore, the performance of the control strategy regarding optimization depends on the rule design. The optimization-based strategies are characterized by the fact that control decisions are calculated using optimization models. A further classification can be done based on the optimization method. Some approaches treat the energy management strategy as an optimal control problem. In this case, the optimization is considered as a whole and needs future information about the requested battery power. Due to the necessity of a priori knowledge of the entire driving profile, this approach is limited to offline applications. In fact, the results of this approach establish a benchmark for evaluating the performance of causal strategies. Furthermore, these results can be used to develop near-optimal energy management strategy suitable for real implementations. Namely, the authors in [9] developed a rule-based strategy representing a trade-off between high accuracy and manageable computing time. It consists in using the results obtained from the optimization model to design the control rule. Other approaches treat the energy management strategy as a static optimization problem that is solved based on the actual information only [10].

With the above survey, it is of great interest to develop the energy management strategy for a reconfigurable battery from the point of view of optimization theory. This would ensure an optimal battery system operation and maximize the profits such as efficiency and lifetime.

1.2 Goals and Contributions of the work

This work aims to develop energy management strategies for reconfigurable batteries using the optimization-based approach to balance the battery cells optimally. First, two mathematical models are established. The extracted energy model describes the interrelationship between the number of active cells and each cell's depth of discharge. The balancing energy model describes the interrelationship between the number of active cells and the battery's balancing level. In a second step, different optimization problems are provided using the developed mathematical models. Since the cells are either active or bypassed, the developed optimization models belong to the class of mixed-integer programming, which required specific search algorithms. Therefore, methods of mixed-integer optimization were examined as part of this work. As a result, three energy management strategies are established, namely a noncausal strategy that ensures the global optimum and two causal strategies for local optimizing well-suited for real-time operation. The validation of the developed strategies is done by means of simulations of realistic scenarios using a battery model experimentally validated.

1.3 Structure of the work

Chapter 1 introduces the field of work. Chapter 2 describes a Lithium-ion storage system from a chemistry point of view as well as electrical modeling. Chapter 3 provides the mathematical models of the reconfigurable architecture. An overview of optimization methods is given in Chapter 4, focusing on dynamic optimization, especially the dynamic programming method. Chapter 5 describes the noncausal optimization strategy. Simulation results illustrated in Chapter 6 show how this strategy maximizes the battery's efficiency. Thereafter, the same work is done for the causal strategies described in Chapter 7. The results of these strategies are shown and then compared with the result of the noncausal strategy in Chapter 8 to prove the feasibility of these real-time strategies. Chapter 9 concludes this work.

Chapter 2

Lithium-ion storage system

2.1 Lithium-ion cell

Lithium-ion cell is a known electrochemical energy storage unit since the second half of the 20th century: first, as non-rechargeable primary cell developed, and then after exhaustive research and development, it has become the most used rechargeable battery.

2.1.1 Cell structure

An electrochemical process takes place in the Lithium-ion cell transforming the chemically stored energy into electrical energy and vice versa. A schematic structure of a Lithium-ion cell is illustrated in figure 2.1. During the charging process, i.e., the electrical energy is converted into chemical energy, the Lithium ions migrate through the separator and intercalate into the graphite. While the loss of electrons oxidizes the active material of the positive electrode, the negative electrode's active material is reduced by the gain of these electrons. During the discharging process, it will come to the revision of the charging process. Lithium atoms release electrons and leave the graphite. The resulting ions move to the positive electrode diffusing through the separator. To establish a charge balance, electrons are released into the

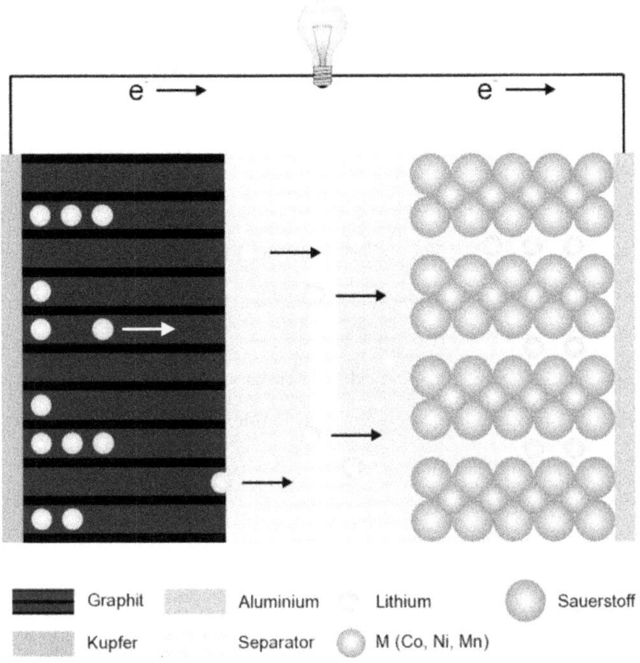

Figure 2.1: Schematic illustration of a Lithium-ion cell during discharging [11] [12].

electric load circuit.

The electrodes of a Lithium-ion cell can be made of different active materials or different additives. The electrochemical potential between both electrodes shall be maximized. The separator separates the two electrodes from each other in order to prevent an intern short circuit.

Various chemical compositions of the cathode have been proposed in the literature, such as Nickel Manganese Cobalt Oxide, Manganese Oxide, and Iron Phosphate. However, an exhaustive investigation of overall factors affecting the safety and the performance of the new material when used in a large-scale battery should occur before its adaption in electric vehicles. For more information about this filed the readers may refer to [13], [12], [14] and [11]. Some of the most common Lithium-ion cell used in electric-vehicle batteries are the Lithium Nickel Manganese Cobalt Oxide and Lithium Iron Phosphate. Each technology has slightly different chemical and electric characteristics [15] [16].

The advantage of a Lithium-ion cell over other storage systems is that the charge and discharge processes are completely reversible and thus are not subject to the memory effect. The redox reaction inside the cell over the discharge and charging process can be divided into an anode and cathode reaction. While the anode reaction is given by

$$\mathrm{Li}^+ + x\ \mathrm{e}^- + \mathrm{C}_6 \rightleftharpoons \mathrm{Li}_x\mathrm{C}_6,$$

a general reaction equation for the charging or discharge process at the cathode can be described as follows [11]

$$\mathrm{Li}_{x+y}\mathrm{MO}_2 \rightleftharpoons x\ \mathrm{Li}^+ + x\ \mathrm{e}^- + \mathrm{Li}_y\mathrm{MO}_2.$$

The chemical reactions inside the cell can be generally described as the diffusion of Lithium ions between anode and cathode through the electrolyte. This chemical reaction involves the transfer of electrons as chemical bonds are formed: Lithium always binds to the negative electrode's graphite layers and the oxidized metals at the positive electrode. The exact number of electrons, Lithium ions, and carbon atoms involved in the reaction depends on the cell's chemical composition.

2.1.2 Cell modeling

Modeling the behavior of a Lithium-ion cell is complex due to non-linear effects and transient during operation. Different models such as electrochemical, analytic, stochastic, and electric models are reported in the literature. These different approaches arise from a variety of application fields. In addition to the improvement of cell performance, the use of an accurate cell model plays an important role in the optimization of the whole battery system. The choice of the convenient model depends on the cell characteristics that need to be modeled and the available computing hardware.

Black-box model

This model describes the electrical behavior of a cell without reproducing the underlying physical processes. The model parameters that do not have physical significance are identified using measurement data describing the cell behavior [17]. This modeling approach is mainly used in the analysis of stationary load use cases. A black-box model has high accuracy in the measured operating points domain, but interpolation or extrapolation can occur in a limited way [18]. An example of a black-box model is the neural network model.

Electrochemical model

The purpose of the physical-chemical model is to describe the cell's internal state in its very detail. Electrochemical and thermodynamic laws are used for the development of the model [19]. In order to model the overall characteristics of the cell, each effect is represented individually by a partial differential equation. These effects are superimposed afterward. Such an approach is known for its high accuracy and the ability to model the cell's dynamic behavior. On the other hand, the computational complexity and parameterization effort are high. This approach is therefore mainly used in the development and optimization of the electrochemical cell.

Equivalent circuit model

This modeling approach is based on the dynamic electrical characteristics of a cell. Such a model uses electrical circuit elements in order to describe the electrical transfer function of a cell in a wide frequency band. Diverse equivalent circuit models are developed, varying from simple model to more detailed model that features the self-discharge rate and the electrochemical overpotentials [20]. The simplest model consists of an ideal voltage source and a resistance connected in series, as shown in figure 2.2. Thus, this equivalent circuit only models the instantaneous cell response to a change in its current.

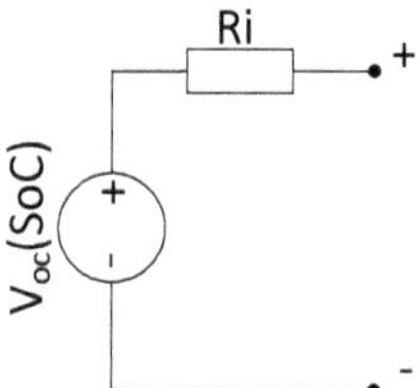

Figure 2.2: Static equivalent circuit model: the cell is modeled as an ideal voltage source with a series resistor.

However, this model does not deal with the dynamics of electrochemistry such as activation and concentration polarization [21] [22]. Thus, the equivalent circuit should be extended by adding one or more RC elements; each is a parallel connection of resistance and capacitor.

Several equivalent circuit models are introduced, which differ in the structure and parameterization effort. The simplest variant is the combination of an ideal voltage source with a serial resistor and a series-connected RC-element, which describes the time dependence of the voltage drop due to the current flow. In order to increase the model accuracy, further RC elements could be added to the equivalent circuit model, as shown in 2.3. Since the degrees of freedom of the model are increased when extending the model, the parameterization effort also increases. For more information about equivalent circuit models, readers are referred to [23] and [24].

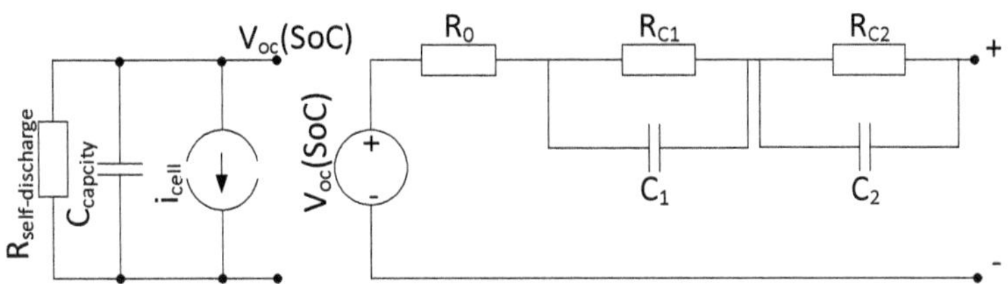

Figure 2.3: Equivalent circuit model to compute the runtime, the current, and the voltage characteristics of a Lithium-ion cell [24].

The Lithium-ion cell's behavior is built by a current source, a capacitor, and a resistor, as shown on the left side of the figure 2.3. The current source gives the cell current released into the load and thus leads to a change of the state of charge of the capacitor representing the cell's capacity. The value of the resistor connected in parallel represents the self-discharge rate which is a function of the state of charge and the temperature. The remainder of the circuit model consists of a voltage source that defines the battery's open-circuit voltage, a series resistor characterizing the ohmic overpotential, and RC elements representing activation and concentration polarization. Thus, the terminal voltage relaxation over time is well described. Namely, effects of the electric double layer and diffusion processes are assigned to RC elements.

The characteristics of the electric elements used in the model are determined using impedance spectroscopy measurements. The elements of the model can be assigned to physical and chemical processes. However, a local consideration of electrochemical or thermodynamic processes is not possible. The advantage of this modeling approach is the low implementation and computing effort, and at the same time, it shows good accuracy. This modeling approach is used in these simulation studies as the goal is to assess the electrical performance of the reconfigurable battery.

2.2 Conventional storage systems

In order to meet customers' expectations regarding power, driving range, and system lifetime, high voltage batteries consist of hundreds of single cells connected in parallel and series, yielding the increase of the battery terminal voltage and the usable capacity, as shown in figure 2.4.

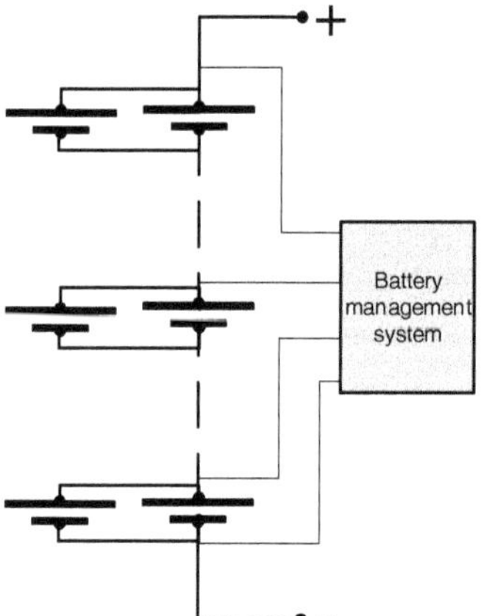

Figure 2.4: Multicell battery having the state of the art architecture: centralized battery management system and static cell connections. Only cells with the same type and the same chemistry could be used.

In order to meet the system requirements, cells are first connected in parallel to form logical cell modules. The parallel connection of cells adds up the capacities of the individual cells, as the current splits in the parallel-connected cells. These logical cells are connected in series afterward. In fact, the individual voltages of the cells are added up in series connection leading to a higher battery voltage.

In addition to capacity increase, the parallel interconnection of many cells in the battery pack averages out the variance of the measured capacity of the individual cell. Consequently, the capacity variance within this battery pack is low compared to the serial interconnection of individual cells with a large capacity. Furthermore, the availability of this pack increases, as a cell failure does not lead to the entire battery pack's failure. However, a disadvantage of connecting many cells of small capacity in parallel is the increased manufacturing cost of such a battery pack.

2.2.1 Energy efficiency issue

The battery efficiency considered in this work is the energy efficiency under discharge that describes to what degree the chemical energy is exploited over a full discharge cycle [25]. Therefore, it is calculated by dividing the amount of electrical energy transferred when discharging $E_{\text{out,elec}}$ by the total energy stored in the battery E_{tot}

$$\eta = \frac{E_{\text{out,elec}}}{E_{\text{tot}}}. \tag{2.1}$$

Note that $E_{\text{out,elec}}$ is given by the integral of the battery current multiplied by the voltage across the battery terminals over the discharge time

$$E_{\text{out,elec}} = \int_{t_0}^{t_e} i_{\text{bat}} \cdot V_{\text{bat}} \, \mathrm{d}t, \tag{2.2}$$

while E_{tot} is defined as follows [25]

$$E_{\text{tot}} = \sum_{j=1}^{N} \int_{Q_{0,j}}^{Q_{e,j}} V_{oc,j}(Q) \, \mathrm{d}Q = \sum_{j=1}^{N} \int_{\xi_{0,j}}^{\xi_{e,j}} V_{oc,j}(\xi) \cdot C_j \, \mathrm{d}\xi, \tag{2.3}$$

where j is the cell index.

The flow of energy during one full discharge cycle can be depicted by a Sankey diagram, as shown in figure 2.5. This diagram summarizes the energy transfers taking place in a discharge process, so self-discharge and capacity fade are not considered.

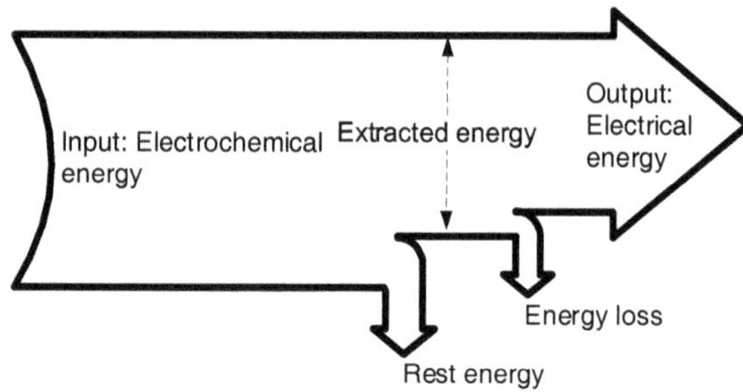

Figure 2.5: This Sankey diagram shows the energy balance for a multicell battery under discharge. The losses (energy loss and rest energy) are shown as an arrow branching out to the bottom.

This Sankey diagram for a multicell battery shows that a part of the input energy is lost in the form of heat. Another part remains unusable when the battery reaches the end-of-discharge. These terms are described in detail below.

Rest energy: is the remaining, not usable energy. As discussed in the introduction, the cell mismatch in a conventional battery limits the battery capacity to the weakest cell. The energy that remains in the battery after reaching cut-off voltage can not be used. Furthermore, in case that the battery has a defective cell, the whole battery is considered damaged and should be recycled [26].

Energy loss: is the amount of the chemical potential energy stored in the battery that is transformed into irreversible heat, which is generated due to three factors. The first factor is the activation overpotential, which describes the kinetic limitation of the charge transfer across the interface between the electrode and the electrolyte.

The second factor is the concentration overpotential. In fact, changes in concentration during the reaction yield a concentration gradient in the cell, resulting in current limitation (mass transfer limitations). The last factor is the ohmic overpotential caused by the pass of electrons through the electrical conductor inside the cell (collectors) [27]. The authors in [28] investigated reversible heating and concluded that this kind of heating could be neglected comparing with irreversible heat in the case of high discharge rate, such as in the battery electric vehicles use case.

2.2.2 Balancing methods

Variations in cell capacities and internal resistances give rise to an unbalanced battery pack. Thus, the available capacity is reduced because the discharging operation is limited by the weakest cell. Therefore, the battery capacity that can be used is expressed as

$$C_{\text{bat}} = \min_{j \in \{1,\ldots,N\}} \{C_1, \ldots, C_N\}, \tag{2.4}$$

where N is the number of cells connected in series and C_{bat} is the battery capacity that can be extracted over a complete discharging cycle.

In addition, in an unbalanced battery pack, not all cells could be fully charged as the battery charging operation is limited by the cell with the best state of charge. Such performance degradation is more significant by storage systems with cells connected in series [2].

In order to address these issues and enhance battery lifetime, cell balancing is required. Figure 2.6 shows how balancing could enhance energy efficiency and fully charge every cell in the battery. As a result, the driving range could be extended.

A review of the literature shows that a lot of research is carried out in the field of balancing methods. These methods are grouped into 2 categories: active balancing and passive balancing. The latter consists of a resistor in parallel to each cell. The role of this resistor is to dissipate remaining energy in the form of heat [29]. This

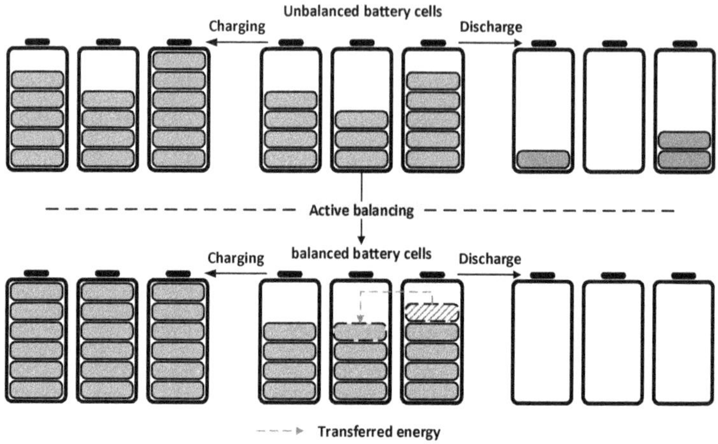

Figure 2.6: The available energy in a multicell battery with cells in series after charging or discharging: unbalanced cells (top), balanced cells (bottom).

allows each cell in the battery to be fully charged. Since passive balancing only converts energy into heat, this process occurs in a predefined operating phase, e.g., during the charging phase. Passive balancing is widely used due to system simplicity and reduced cost.

A lot of studies have been focused on balancing methods where the excess energy is shifted between battery cells instead of dissipating the energy in the form of heat. These solutions called active balancing are more efficient than passive balancing. Active balancing methods can be categorized into the following categories: cell-to-cell, pack to cell, cell to pack, and cell(s) to pack to cell(s). Figure 2.7 lists some examples of active balancing circuits often tackled in the literature.

Basically, active balancing circuits use short-term energy storage devices or conversion elements such as transformers, capacities, and inductors to transfer energy either from one cell to another or between one cell and the whole battery pack. The authors in [31] presented a balancing circuit named switched capacitor. It does

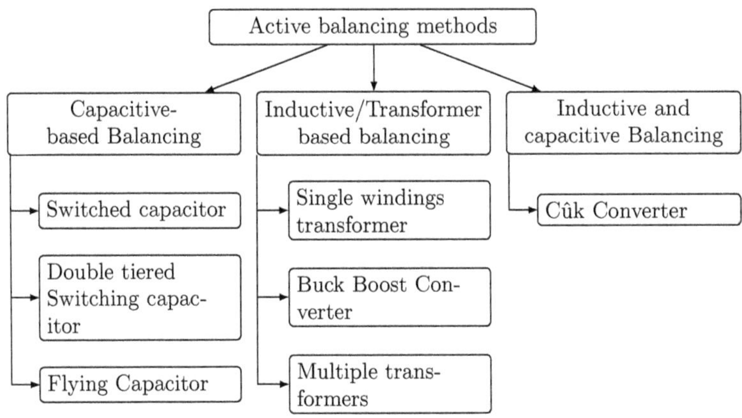

Figure 2.7: Classification of the different active balancing methods according to the used elements [3] and [30].

not need any intelligent control strategy as the switches are frequently moved from two positions (upper and lower). However, the main drawback is the long balancing time. To address this issue, costly circuits with more components and complex control strategies are needed [32]. In [29] [33] a comparison of different balancing structures is carried out. In [34] a state-space model is used to describe the cell balancing circuit. Simulations help by comparing the various balancing methods and optimizing the capacitors and transformer parameters used in the circuit.

In addition, different control approaches for active balancing are investigated. The authors in [35] used a nonlinear model-based predictive control to optimize the strategy of the described active balancing circuit. A genetic algorithm was used to solve the resulting optimization problem.

Instead of using energy storage devices, DC-DC converters are applied to equalize battery cells. Strategies based on this approach are originated in [36] and other research [37] and [38]. For more information about the active balancing circuits, readers are referred to [3] and [30].

Despite higher efficiency than passive balancing, the active balancing method faces several challenges such as limited balancing current, higher cost, packaging size, and control strategy complexity. Thus, these issues make active balancing not attractive for automotive applications. Namely, the main disadvantage of balancing with a limited current is the balancing time. The more significant the cell-to-cell variation, the longer the balancing time is needed, and thus it will be difficult to equalize all battery cells in good time. As a result, the balancing circuit fails to operate as expected, yielding a modest and insufficient improvement of available battery capacity. Consequently, the limited balancing current prohibits the use of cells with unsorted capacities or different nominal capacities connected in series.

Since these balancing circuits are designed for the conventional battery architecture in which the cells are permanently connected to the load, they cannot overcome the failure of the storage system by a defective single cell. To enhance the energy efficiency as well as battery uniformity and lifetime, a modular battery system that consists of cell modules connected in series using semiconductor switches was proposed in the literature as a promising approach towards the disadvantages of active and passive balancing circuits. These semiconductor devices allow continued battery operation by bypassing the defective or weak cell. Such systems will be dealt with in more detail in the next section 2.3.

2.3 Reconfigurable storage systems

To prevent performance degradation or failing battery, the static cell connections are interrupted by switches that enable dynamic cell connection, as shown in figure 2.8.

Such modular battery architecture allows dynamic reconfiguration of the battery pack and controlling the load of each cell. Actually, the battery management system defines, depending on the control strategy, which cells to be bypassed to be disconnected from the current path. Connecting and disconnecting battery cells

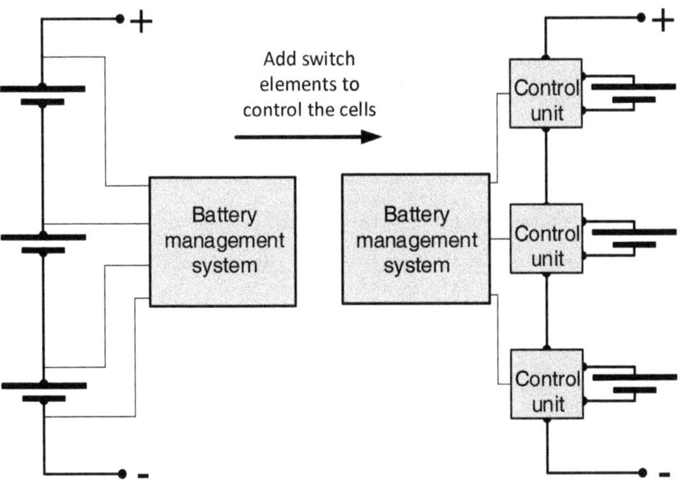

Figure 2.8: From the static setup of cells connected in series and in parallel towards a reconfigurable architecture [39].

appropriately allows an optimal equalization of the cells, leading to enhancing the battery's energy efficiency and lifetime.

2.3.1 Various architecture concepts

Reconfigurable battery systems are widely discussed in the literature from the point of view of hardware realization. Different approaches for arranging switches in the battery are reported. While in [40] five switches are used for each cell, the authors in [41] and [42] proposed up to six switches to control each cell. Both architectures are shown in figure 2.9 (d) and (c), respectively. Such architectures allow connecting cells either in parallel, in series, or a combination of both. Hence, the optimal configuration of the pack can be chosen for different load phases. The authors in [43] and [44] proposed a battery architecture where i cells are connected in parallel, building a logical cell module to provide a higher capacity, and j logical cells are connected in series to achieve a higher terminal battery voltage. The special feature of the presented approach is that each cell is equipped with a separate switch which enables disconnect the cell from the module. The individual module can be bypassed by an additional switch as illustrated in figure 2.9 (f). An iterative algorithm was developed to control the switches to balance the cell modules. In [45] a battery architecture is presented where three switches per cell are used. These switches are controlled independently from each other. This topology corresponds to figure 2.9 (b).

An approach with seven switches per cell is investigated in [46]. This approach is depicted in figure 2.9 (e). It allows splitting the battery into two packs which can be used independently by primary and secondary loads. However, using a large number of switches per cell affects system reliability and increases the system cost. To overcome these issues, only two switches per cell pack are considered in [47] so that the battery can only be reconfigured on the cell pack level. Figure 2.9 (a) illustrates the proposed architecture.

The drawback of using switches on the cell pack level is that all cells of the pack are active or bypassed at the same time regardless of the cell mismatch within the pack. As a consequence, an optimal balancing of each cell is not possible in this case. The effect of switch module size on energy efficiency was investigated in [48] using Monte-Carlo simulations. These simulations have shown that the energy storage system's efficiency decreases by more than 10% when using one switch module for 10 cells.

2.3.2 Proposed reconfigurable battery architecture

The reconfigurable battery architecture considered in this work is based on the energy storage proposed in [24], [49], [50], and [4]. In these studies, an architecture that consists of connecting only two switches to each cell was introduced. This architecture allows to connect or bypass the single battery cell. In [4], a concept

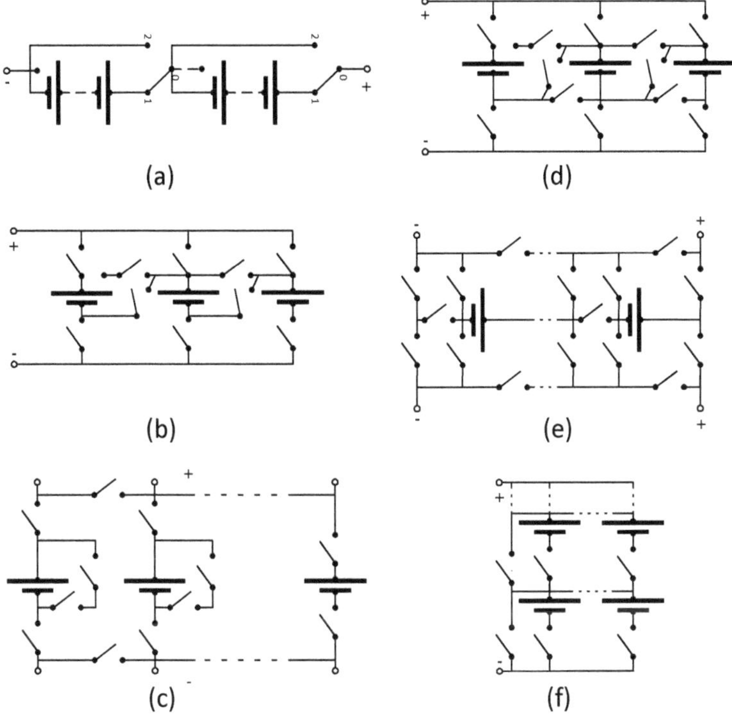

Figure 2.9: Different architectures of reconfigurable batteries. (a) the reconfigurable topology on cell pack level. (b), (c), (d), (e), and (f) are reconfigurable architectures on cell level.

for applications such as the electric bicycle was suggested. In contrast to the architectures with three or more switches, the cell connections in the proposed battery architecture can only be reconfigured in series. So a parallel configuration or combination of series and parallel is not provided. The advantage of this architecture is that it represents a trade-off between cost and flexibility. Figure 2.10 illustrates this battery architecture.

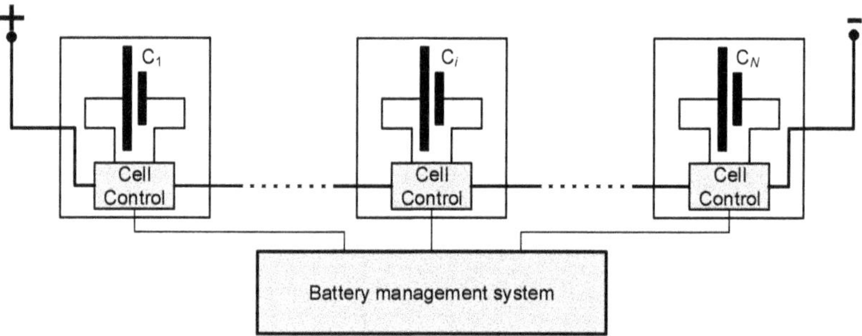

Figure 2.10: Architecture of the proposed reconfigurable battery. It includes a battery management system and cell control units enable to connect or disconnect each cell in the battery.

As can be seen in figure 2.10, battery cells are decoupled and do not have a direct connection between each other. Instead, each cell is connected to a power control unit forming the cell module. These modules are connected in series to meet the rated voltage required. The battery management system classifies the cells according to their actual state of charge and selects the best cells based on the implemented control strategy.

The technical feasibility of the reconfigurable topology was the subject of many research. The authors in [51] focused on the characteristics of power electronic switches for automotive applications. They conducted a review of power electronic switches and their properties regarding power dissipation, rating current, and breakdown voltage. These factors represent crucial features for the decision of whether the power electronic switch can be implemented in the electric vehicle battery or not. Figure 2.11 shows the structure of four types of power devices-

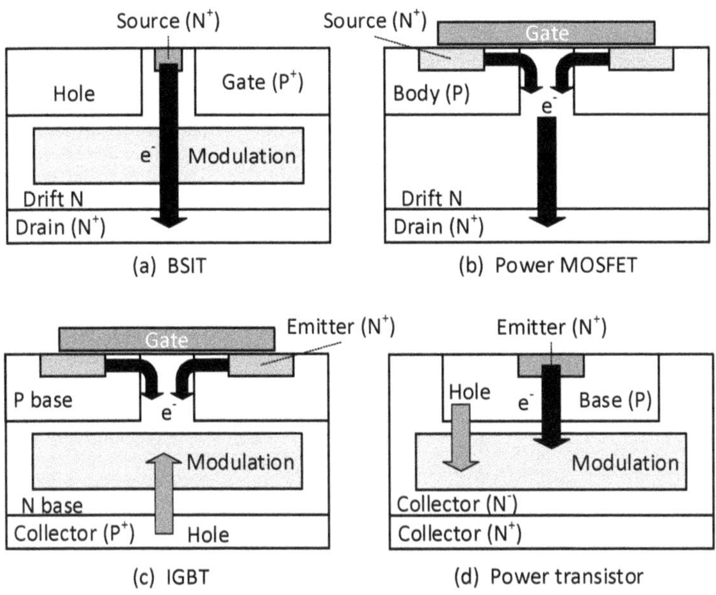

Figure 2.11: Schematic representation of the power devices structure: (a) Bipolar mode static induction transistor (BSIT), (b) Power MOSFET, (c) Insulated gate bipolar transistor (IGBT), and (d) Power transistor. While gray arrows show the hole flow, Black arrows indicate the flow of electrons [51].

The power MOSFET has the advantage of being able to switch large currents. In fact, the ON resistance between drain and source R_{DSon} is very small, leading to reduced power dissipation. The authors in [47] conducted thermal experiments verifying the MOSFET capability to sustain high current values, such as by the electric vehicles. Furthermore, a gate voltage input controls the current flow between drain and source. While a positive voltage should be applied to the gate to saturate the MOSFET, zero volts turn the switch off, so a simple control circuit can switch the device. Additionally, MOSFET has very high input (Gate) resistance making it possible to safely connect many different MOSFETs in parallel to achieve the current required. The features mentioned above make the MOSFET a suitable switch to connect or disconnect battery cells during operation. Figure 2.12 illustrates the schematic of the proposed cell module including two switches which are bypass and current path MOSFETs.

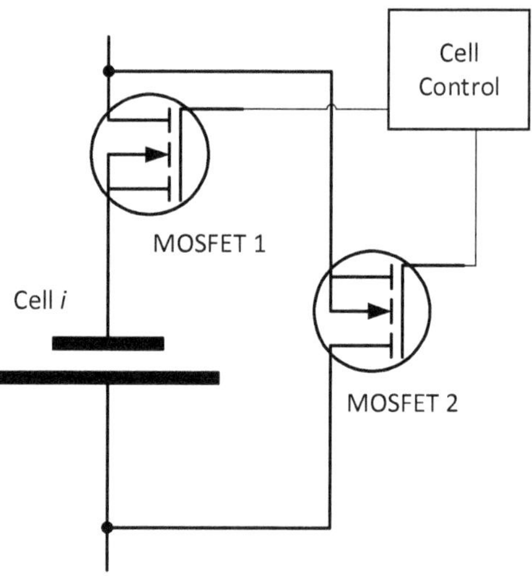

Figure 2.12: Schematic of a switch module for a cell in a reconfigurable battery using n-channel power MOSFETs.

2.3.3 Control strategies for two switch per cell architecture

Reconfigurable battery architecture allows an optimal operation of the battery. Besides the prevention of overcharging or deep-discharging of battery cells, it is possible to equalize cells efficiently. Switches are placed between the individual cells to connect or disconnect these cells. Dynamically switching between cells allows balancing the battery. Namely, a bypassed cell does not provide current as long as it is disconnected, whereas the activated cells are discharged. Hence, this is a new approach compared to the passive and active balancing methods; it enables equalizing the different states of charge of cells through dynamic switching between selected cells. In order to operate the switch modules, an energy management strategy is needed. Which cells should be bypassed depends on the implemented strategy's objective, e.g., maximizing the electrical energy transferred when discharging or minimizing the battery's rest energy. Actually, this energy management strategy can be developed using different approaches. As shown in figure 2.13, these approaches can be clustered into optimization-based and rule-based approaches [6], [8], and [52].

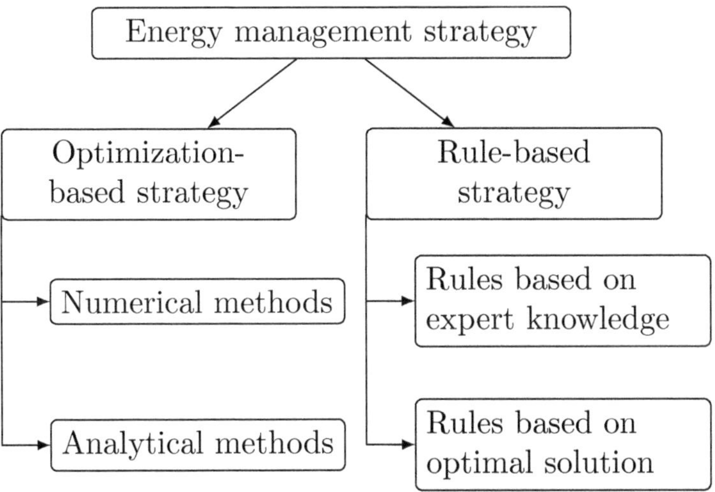

Figure 2.13: Classification of energy management strategies according to the approach used to develop the strategy.

An online optimization during operation characterizes the optimization-based approach. In contrast, the decisions by rule-based approach are made based on previously defined rules. Note that the rule-based strategies differ in the design of the control laws. The range extends from relatively simple rules based on intuition to complex rules definition.

The focus of the most previous research was the optimization of the hardware design. Concerning control strategies, the published research almost implemented rule-based strategies. In [4] a balancing algorithm with two possible control variables, i.e., cell voltage or state of charge, was introduced. This algorithm classifies the battery cells based on either their terminal voltage or their state of charge. However, the number of cells that could be disconnected is fixed regardless of the cell spread within the battery or the power request. Namely, the deviation of the control variable value from the mean value is calculated for each cell. Then, during the discharge phase, only the cell with the lowest deviation is bypassed. In the charging phase, however, only the cell with the highest deviation is bypassed. The term *one-cell-off strategy* is used throughout this work to denote this strategy.

Unlike optimization-based control strategies, decisions are made based on predefined rules. The rules are usually implemented as state machines or in the form of characteristic maps. These maps describe the decisions as a function of various input factors. An approach that relies on fuzzy logic is presented in [53]. The complexity of rules design varies from relatively simple approaches based on intuition and simple relationships, approaches based on efficiency analyses up to approaches with rules derived from optimization-based solutions.

The optimization-based operating strategies are not popular in the field of reconfigurable batteries. However, in the domain of hybrid vehicles, several optimization-based approaches have been developed in recent years. Authors in [54] used optimization techniques such as linear programming and optimal control. These models offer a globally optimal solution. As they are not suitable for online implementation, their result can be used to derive complex rules for rule-based strategies representing a trade-off between complexity and accuracy. Moreover, these noncausal strategies serve to benchmark the performance of other causal strategies which only consider the current state. Therefore, they provide a local optimal solution. Such causal strategies are presented in [55].

Actually, an optimization-based energy management strategy can also be used in the battery management system of the reconfigurable battery proposed in this work. In fact, the control strategy shall switch the cells based on the result of an optimization problem so that at any time, the optimal module configuration provides the power demand. In contrast to previous research, this work focuses on developing optimal energy management strategies for reconfigurable batteries. The dynamic optimization techniques are used to maximize the energy management strategy. The optimization problems will be discussed in detail in chapters 5 and 7.

Chapter 3

Model of reconfigurable battery

This chapter introduces the mathematical modeling of the reconfigurable topology considered in this work. First, an analysis of the impact of the number of active cells on the battery cells' depth of discharge is provided. Thereafter, a mathematical model is established that describes the balancing energy as a function of the number of active cells. Finally, an energy loss model is formulated to calculate the irreversible heat generated during operation.

3.1 Definition of the cell switching process

In contrast to conventional storage systems, the cells within the reconfigurable battery are temporarily bypassed based on their actual energy content; in the way that only the best n_{on} cells are selected to provide the current to the load. By disconnecting n_{off} cells, the connected cells should deliver more electrical charges than when all battery cells N would be connected. The amount of energy that is additionally drawn from the active cells over a time interval allows the cells' equalization during battery operation.

The purpose of cell equalization is to enhance battery efficiency by minimizing the unused energy remaining in the battery at the end of discharge. The equalization level indicates how far cells have a similar state of charge. The equalization level reached at the end of the discharge depends on the initial cell spread within the battery and the number of cells active during operation. In fact, if a large number of cells is activated, the cells with a low state of charge are fully discharged too quickly, resulting in incompletely balanced cells with remaining unused electrical energy. On the other hand, the larger the number of bypassed cells, the faster the cells are balanced, and the less energy remains in the battery after reaching the end of discharge criteria. Hence, the equalization level influences the amount of energy that can be extracted. At first sight, the number of active cells is assumed to be constant during discharge. n_{on} and n_{off} describe the number of active cells and the number of bypassed cells, respectively. In the next section, more insight will be given for the dependency of the remaining energy at the end of discharge on the number of active cells.

3.2 Operation state of the battery cells

Let N denote the total number of cells in the battery and $B = \{C_1, C_2, \ldots, C_N\}$ the set of the cell capacities. Note the index $j \in \{1, 2, \ldots, N\}$ is used to identify the cells in the battery.

Cell switches used in the reconfigurable battery are mathematically described by the operation state vector $\boldsymbol{s}$ whose components $\{s_1, s_2, \ldots, s_N\}$ indicate the operation state (connected or bypassed) of the battery cells, respectively. Hence, $\boldsymbol{s} \in \{0, 1\}^{\text{N}}$ is a binary vector that defines which cells are connected, i.e., $s_j = 1$, and which are disconnected, i.e., $s_j = 0$

$$s_j = \begin{cases} 1 & \text{if the cell } j \text{ is active} \\ 0 & \text{if the cell } j \text{ is disabled} \end{cases} . \tag{3.1}$$

Let $\boldsymbol{i}_{\textbf{cell}}$ denote the vector

$$\boldsymbol{i}_{\textbf{cell}} = \begin{pmatrix} i_1 \\ \vdots \\ i_N \end{pmatrix} \tag{3.2}$$

that includes the current values i_j of battery cells. In a series circuit, the amount of current is the same through any connected cell. Therefore,

$$\boldsymbol{i}_{\text{cell}} = i_{\text{bat}} \cdot \boldsymbol{s} = i_{\text{bat}} \cdot \begin{pmatrix} s_1 \\ \vdots \\ s_N \end{pmatrix}, \tag{3.3}$$

where vector $\boldsymbol{s}$ includes the operation states of the battery cells and i_{bat} is the algebraic value of the battery current.

3.3 Extracted energy

After modeling the operational state of a cell in a reconfigurable battery, the energy extracted from the battery can be expressed as a function of the number of active cells.

Let $Q_{\text{th},j}(t)$ denote the cell charge throughput, which is the accumulated charge that the cell j provided during the time interval $[0, t]$:

$$Q_{\text{th},j}(t) = \int_0^t i_j(\tau)\,\mathrm{d}\tau. \tag{3.4}$$

Sum both sides of the equation (3.4) from $j = 1$ to $j = N$

$$\sum_{j=1}^{N} Q_{\text{th},j}(t) = \sum_{j=1}^{N} \int_0^t i_j(\tau)\,\mathrm{d}\tau. \tag{3.5}$$

The fact that the sum of integral is the integral of sum leads to the following

$$\sum_{j=1}^{N} Q_{\text{th},j}(t) = \int_0^t \sum_{j=1}^{N} i_j(\tau)\,\mathrm{d}\tau. \tag{3.6}$$

Looking back to the equation 3.3, $\sum_{j=1}^{N} Q_{\text{th},j}(t)$ can be expressed as follows

$$\sum_{j=1}^{N} Q_{\text{th},j}(t) = \int_0^t \sum_{j=1}^{N} s_j(\tau) \cdot i_{\text{bat}}(\tau)\,\mathrm{d}\tau. \tag{3.7}$$

Remember that s_j with $j \in \{1, 2, \ldots, N\}$ indicates the status of the cell j either is connected or bypassed. Thus, the sum of all vector elements s_j represents the number of active cells n_{on}. Assuming that the number of active cells is constant during battery operation, the sum $\sum_{j=1}^{N} s_j = n_{\text{on}}$ does not depend on the time. $\sum_{j=1}^{N} Q_{\text{th},j}(t)$ thus be calculated according to the following

$$\sum_{j=1}^{N} Q_{\text{th},j}(t) = \sum_{j=1}^{N} s_j(\tau) \cdot \int_0^t i_{\text{bat}}(\tau)\,\mathrm{d}\tau. \tag{3.8}$$

Substituting the value of $\sum_{j=1}^{N} s_j(\tau)$ into equation 3.8 leads to the following

$$\sum_{j=1}^{N} Q_{\text{th},j}(t) = n_{\text{on}} \cdot \int_0^t i_{\text{bat}}(\tau)\,\mathrm{d}\tau. \tag{3.9}$$

Let $Q_{\text{th,bat}}(t)$ denote the battery charge throughput, which is the accumulated charge that the battery provided during the time interval $[0, t]$

$$Q_{\text{th,bat}}(t) = \int_0^t i_{\text{bat}}(\tau)\,\mathrm{d}\tau. \tag{3.10}$$

As a result, the relation between the number of active cells and the accumulated charge extracted from the battery can be described by a mathematical formula as follows

$$n_{\text{on}} = \frac{\sum_{j=1}^{N} Q_{\text{th},j}(t)}{Q_{\text{th,bat}}(t)}. \tag{3.11}$$

The cut-off criteria: In this paragraph, the end of discharge criteria of a reconfigurable battery is introduced. Remember that the purpose is to discharge the battery cells so that they reach the cut-off voltage almost simultaneously. That means that all battery cells show a low voltage level at the end of the operation. Therefore, further operating the battery after disconnecting the empty cells bears the danger of damage the active cells. As a result, the end of discharge criteria is defined as the time t_{e} at which the first cell or group of cells reach the cut-off voltage.

Let $\boldsymbol{m} \in \{0, 1\}^N$ be a binary vector that indicates which of the battery cells are fully discharged at time t_{e}. Namely, $m_j = 1$, if the accumulated charge $Q_{\text{th},j}(t_{\text{e}})$ of the cell j is equal to the amount of charge stored in this cell at the beginning of discharge $Q_{0,j}$. Whereas, $m_j = 0$ if the accumulated charge $Q_{\text{th},j}(t_{\text{e}})$ of the cell j is smaller than the amount of charge stored in this cell at the beginning of discharge $Q_{0,j}$. Hence, the elements of vector $\boldsymbol{m}$ are defined according to the following

$$m_j = \begin{cases} 1 & \text{if } Q_{\mathrm{th},j}(t_\mathrm{e}) = Q_{0,j} \\ 0 & \text{if } Q_{\mathrm{th},j}(t_\mathrm{e}) < Q_{0,j}. \end{cases} \tag{3.12}$$

Let M denote the number of fully discharged cells at time t_e, i.e., cells with 100% depth of discharge. Thus, M can be calculated as follows

$$M = \sum_{j=1}^{N} m_j. \tag{3.13}$$

The control strategy continually computes the n_on best cells and connects them to the load. When the battery cells are close to each other regarding the state of charge, continuing changes of cells order every sample time. Therefore, the control strategy constantly switches between these cells. As the switching frequency is much higher than the cell dynamic (state of charge decrease), the cells involved in the switching process reach the end of discharge voltage simultaneously. Hence, cells, which are not fully discharged at time t_e, are not switched off over time. Thus, the charge throughput of each cell is given as

$$Q_{\mathrm{th},j}\left(t_\mathrm{e}\right) = Q_{\mathrm{th,bat}}(t_\mathrm{e}), \quad \text{for } j \in \{j \in \{1,\ldots,N\} | m_j = 0\}. \tag{3.14}$$

Looking back at the equation (3.11) express the battery charge throughput at time t_e is expressed as

$$Q_{\mathrm{th,bat}}(t_\mathrm{e}) = \frac{\sum_{j=1}^{N} Q_{\mathrm{th},j}(t_\mathrm{e})}{n_\mathrm{on}}. \tag{3.15}$$

The numerator in equation 3.15 represents the sum of charge throughputs of battery cells. Actually, the charge throughput $Q_{\mathrm{th},j}$ of the cell j at time t_e is equal to $Q_{0,j}$ if $Q_{\mathrm{th,bat}}(t_\mathrm{e}) > Q_{0,j}$; otherwise, $Q_{\mathrm{th},j}$ is equal to $Q_{\mathrm{th,bat}}(t_\mathrm{e})$. Using the aforementioned mapping, the sum of charge throughputs of battery cells is given by the following expression

$$\sum_{j=1}^{N} Q_{\text{th},j}(t_\text{e}) = \sum_{j=1}^{N} Q_{0,j} \cdot m_j + \sum_{j=1}^{N} Q_{\text{th,bat}}(t_\text{e}) \cdot (1 - m_j). \tag{3.16}$$

The first term on the right in the equation (3.16) represents the sum of charge throughput of battery cells not fully discharged at time t_e. These cells have the same charge throughput $Q_{\text{th,bat}}(t_\text{e})$. Therefore, the following relation is obtained

$$\sum_{j=1}^{N} Q_{\text{th,bat}}(t_\text{e}) \cdot (1 - m_j) = N \cdot Q_{\text{th,bat}}(t_\text{e}) + Q_{\text{th,bat}}(t_\text{e}) \cdot \sum_{j=1}^{N} m_j. \tag{3.17}$$

Substituting the value of $\sum_{j=1}^{N} m_j$ into equation 3.17, the battery charge throughput $Q_{\text{th,bat}}(t_\text{e})$ results in

$$\sum_{j=1}^{N} Q_{\text{th,bat}}(t_\text{e}) \cdot (1 - m_j) = N \cdot Q_{\text{th,bat}}(t_\text{e}) + M \cdot Q_{\text{th,bat}}(t_\text{e}). \tag{3.18}$$

Thus, equation 3.16 can be expressed as follows

$$\sum_{j=1}^{N} Q_{\text{th},j}(t_\text{e}) = \sum_{j=1}^{N} C_j \cdot \xi_{0,j} \cdot m_j + (N - M) \cdot Q_{\text{th,bat}}(t_\text{e}). \tag{3.19}$$

Note that C_j is the capacity of cell j and $\xi_{0,j}$ its initial state of charge. The identity $Q_{0,j} = C_j \cdot \xi_{0,j}$ is valid under the assumption that the cell capacity C_j represents the effective usable capacity.

Substituting the value of $\sum_{j=1}^{N} Q_{\text{th},j}(t_\text{e})$ into equation 3.15, the battery charge throughput $Q_{\text{th,bat}}(t_\text{e})$ is expressed as

$$Q_{\text{th,bat}}(t_\text{e}) = \frac{\sum_{j=1}^{N} C_j \cdot \xi_{0,j} \cdot m_j + (N - M) \cdot Q_{\text{th,bat}}(t_\text{e})}{n_\text{on}}. \tag{3.20}$$

By evaluating equation 3.20, the following relation is obtained

$$Q_{\text{th,bat}}(t_\text{e}) = \frac{\sum_{j=1}^{N} C_j \cdot \xi_{0,j} \cdot m_j}{n_\text{on} - (N - M)}, \quad \text{for } n_\text{on} \neq (N - M). \tag{3.21}$$

Equation 3.21 describes the relation between n_on, m_j, and $Q_{\text{th,bat}}(t_\text{e})$. As the vector $\boldsymbol{m} = (m_1, \ldots, m_N)^\intercal$ indicates which battery cells have 100% depth of discharge when the cut-off criterion is reached, the elements $(m_1, \ldots, m_N)^\intercal$ can be determined according to the value of M. For instance, $M = 1$ means only one cell is fully discharged, which is the cell with the smallest initial charge Q_0. This results in

$$m_j = \begin{cases} 1 & \text{for } j = \arg\min_{j \in \mathbb{N}}\{Q_{0,1}, \ldots, Q_{0,j}, \ldots, Q_{0,N}\} \\ 0 & \text{otherwise.} \end{cases} \tag{3.22}$$

Analogously, for $M = 2$ the vector $\boldsymbol{m}$ has only two non-zero elements m_p and m_q, where p and q are the indices of the two cells with the two initial charge values. The same is valid for $M = 3$ until $M = N$. Once the vector $\boldsymbol{m}$ is determined for each $M \in \{1, \ldots, N\}$, the sum of charge throughput of fully discharged cells $\sum_{j=1}^{N} C_j \cdot \xi_{0,j} \cdot m_j$ can be calculated. As a next step, using the equation 3.21, the battery charge throughput $Q_{\text{th,bat}}$ is determined for $n_{\text{on}} \in \{1, \ldots, N\}$.

Furthermore, the following facts should be taken into consideration. First, each battery cell whose charge throughput smaller than the battery charge throughput is fully discharged. Analogously, each battery cell whose initial charge $Q_{0,j}$ is larger than the battery charge throughput is still not depleted. Therefore, to prevent unfeasible solutions, the following restriction is introduced

$$m_j = \begin{cases} 1 & \text{if } Q_{\text{th,bat}}(t_e) \geq C_j \cdot \xi_{0,j} \\ 0 & \text{if } Q_{\text{th,bat}}(t_e) < C_j \cdot \xi_{0,j}. \end{cases} \tag{3.23}$$

As a result, using the equations 3.21, 3.23, and 3.13, the following system of equations is built

$$\begin{cases} Q_{\text{th,bat}}(t_e) = \dfrac{\sum_{j=1}^{N} C_j \cdot \xi_{0,j} \cdot m_j}{n_{\text{on}} - (N - M)} \\ M = \sum_{j=1}^{N} m_j \\ m_j = 1 & \text{if } Q_{\text{th,bat}}(t_e) \geq C_j \cdot \xi_{0,j} \\ m_j = 0 & \text{if } Q_{\text{th,bat}}(t_e) < C_j \cdot \xi_{0,j}. \end{cases} \tag{3.24}$$

Equation 3.24 allows the calculation of the battery charge throughput $Q_{\text{th,bat}}(t_e)$ and the number of fully discharged cells M as a function of the number of active cells.

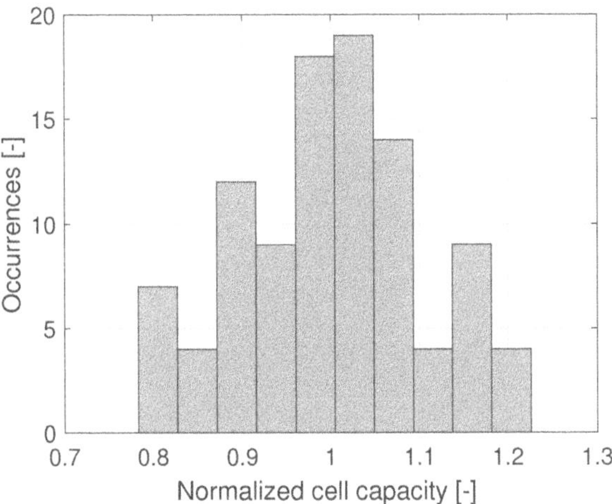

Figure 3.1: Normally distributed cell capacities are randomly generated (capacities are normalized to the mean capacity). The capacity variance is 5.3%.

For the battery shown in figure 3.1, the relationship between the number of active cells and the charge throughput is illustrated in figure 3.2. It is clear that the fewer cells are active, the larger is the accumulated charge that the battery provided during the discharge.

Once the system of equations (3.24) is solved and the battery charge throughput is known, the extracted energy at the end of discharge can now be calculated. As it is the amount of energy left in the battery during the operation, it can be obtained from the following expression

$$E_{\text{ext}} = \sum_{j=1}^{N} E_j \cdot m_j + \left| \sum_{j=1}^{N} \int_{\xi_{0,j}}^{\xi_{e,j}} V_{oc,j}(\xi) \cdot C_j \cdot (1 - m_j)\, \mathrm{d}\xi \right|, \tag{3.25}$$

where E_j is the total cell energy and $\xi_{e,j}$ is the state of charge of the cell j at the end of discharge.

Using the aforementioned equations (3.25) and (3.24), the function $f_E(\cdot)$ is defined

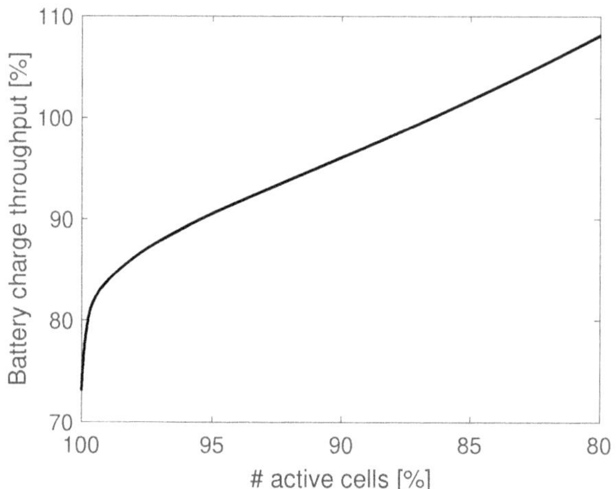

Figure 3.2: The accumulated charge (normalized to the biggest nominal cell capacity of the battery) provided by a reconfigurable battery over a full discharge cycle as a function of the number of active cells (normalized to the total number of battery cells). The simulated battery is shown in figure 3.1.

to map the number of active cells $n_{\text{on}} \in \chi = \{1, \ldots, N\}$ to the energy extracted during discharge E_{ext}. This function is defined on the set χ, called domain, and assigns exactly one value from $\mathbb{R}$, called codomain, to every element from the domain

$$\begin{aligned} f_E \colon \ \chi \subset \mathbb{N} \quad &\rightarrow \quad \mathbb{R} \\ n_{\text{on}} \quad &\rightarrow \quad E_{\text{ext}}. \end{aligned} \tag{3.26}$$

Figure 3.3 shows the graph of the function $f_E(\cdot)$ applied to the battery shown in figure 3.1 As shown in figure 3.3 the evolution of the extracted energy E_{ext} can be analytically described by the function $f_E(\cdot)$. This function maps each number of active cells during operation defined in the input domain to exactly one output value $f_E(n_{\text{on}})$, which represents the energy extracted from the battery until the cut-off criterion is reached.

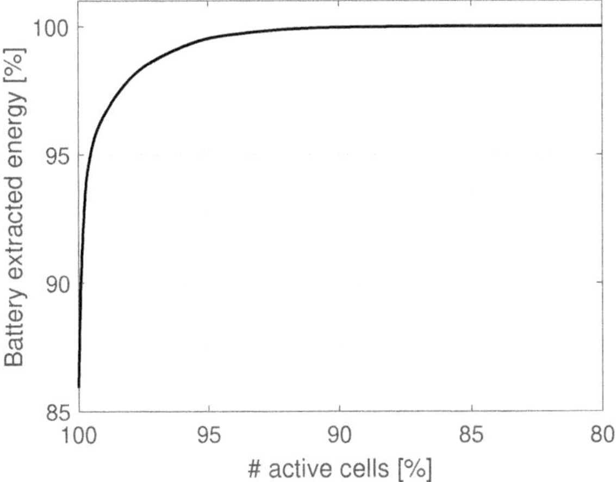

Figure 3.3: The energy extracted from the battery at the end of a full discharge cycle as a function of the number of active cells (normalized to the total number of battery cells). The simulated battery is shown in figure 3.1.

In conclusion, the relationship between the number of active cells n_{on} and the extracted energy E_{ext} have been successfully described by means of the mathematical formula (3.25). Using this correlation, the function $f_E(\cdot)$ mapped each n_{on} from the domain of definition χ to an extracted energy E_{ext}.

3.4 Balancing energy

As mentioned in the previous section, the equalization of the cells during battery operation can be achieved due to cell switching. In this paragraph, it will be explained how cells can be balanced in a reconfigurable battery.

In fact, the functional principle of the reconfigurable battery proposed in this work consists of always connecting the n_{on} best cells with respect to state of charge to the load. As the remaining battery cells are bypassed, they do not provide current.

Hence, the state of charge of every bypassed cell is frozen while the state of charge of connected cells decreases. Hence, bypassing cells over the discharge process forces the connected cells to provide more electrical charges than in case of all battery cells would be active and provide power. This effect is depicted in figure 3.4.

As the power required only depends on the load, the energy additionally provided by the active cells over a time interval Δt corresponds to the electrical energy that bypassed cells would have delivered over the same time interval Δt. Therefore, this amount of energy is called equalization energy and denoted by E_{bcg}, which can be calculated according to the following formula

$$\sum_{j=1}^{N} \Delta E_j - \sum_{j=1}^{N} (\Delta E_j + \Delta E_{\mathrm{bcg},j}) \cdot s_j = 0, \tag{3.27}$$

where ΔE_j is the energy provided by the cell j over Δt in the case all battery cells would be active and $\Delta E_{\mathrm{bcg},j}$ is the equalization energy provided by the active cell j over Δt. Rearranging equation (3.27) so that the equalization energy over Δt is the subject

$$\sum_{j=1}^{N} \Delta E_{\mathrm{bcg},j} \cdot s_j = \sum_{j=1}^{N} \Delta E_j - \sum_{j=1}^{N} \Delta E_j \cdot s_j. \tag{3.28}$$

Considering the entire discharge time interval, i.e., Δt is the time needed to discharge the battery completely, the following mathematical formula describes the achieved equalization energy E_{bcg}

$$E_{\mathrm{bcg}} = \sum_{j=1}^{N} (1 - s_j) \cdot E_j. \tag{3.29}$$

The right-hand side of equation (3.29) represents the energy content of the cells supposed to be bypassed.

3.5 Energy loss

According to the approach by the authors in [10] and [56], each cell is represented by a static equivalent circuit model, which comprises a voltage source connected

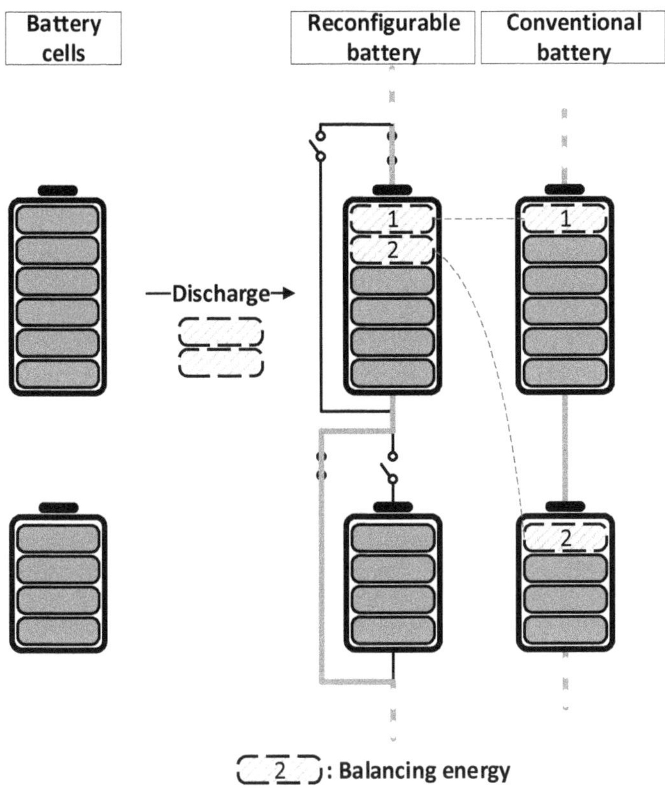

Figure 3.4: Schematic illustration of the balancing process within a reconfigurable battery. The energy provided by the bypassed cell in the case both cells would be active corresponds to the Balancing energy extracted from the active cell.

in series with a resistor, as shown in figure 2.2. This resistor Ri represents the cell internal resistance which is assumed to be constant. The value of the internal resistance Ri is defined in such a way that the voltage drop does not only depend on the ohmic resistance R_0 but also the dynamic effects are taken into account, i.e., electrochemical polarization and the concentration polarization R_p. In fact, the cell internal resistance is determined based on the voltage drop after 10 seconds for charging or discharge with 1C, as shown in figure 3.5.

On the other hand, in the dynamic model, the internal resistance value is determined based on the voltage drop within a short time, e.g., 0.1 seconds. This is due to the fact that the activation- and concentration- overpotentials in the dynamic model are modeled by the RC elements. In contrast to the 10 seconds value, the 0.1 seconds value contains almost exclusively the ohmic resistance of the internal resistor Ri. While this value comes from a battery's measurements, the RC elements are parameterized by comparing the simulated battery voltage with the values measured in different driving profiles [52].

Using this model shown in figure 2.2, the goal is to determine the relationship between energy loss and the number of active cells. Applying the second Kirchhoff's law to the electrical circuit depicted in figure 2.2, the cell current i_j is given by the formula

$$i_j = \frac{V_{\text{oc},j} - V_j}{Ri_j} \cdot m_j \quad \text{for } j \in 1, \ldots, N, \tag{3.30}$$

where $V_{\text{oc},j}$ is the cell open-circuit voltage, V_j the cell terminal voltage, and Ri_j is the cell internal resistance. By multiplying both sides by i_j results in

$$i_j^2 = \frac{V_{\text{oc},j} \cdot i_j - P_j}{Ri_j} \cdot m_j \quad \text{for } j \in 1, \ldots, N, \tag{3.31}$$

where P_j denotes the cell power. By evaluating the quadratic equation 3.31, the cell current i_j is expressed as follows

$$i_j = \frac{V_{\text{oc}} - \sqrt{V_{\text{oc}}^2 - 4 \cdot Ri_j \cdot P_j}}{2 \cdot Ri_j} \cdot m_j \quad \text{for } j \in 1, \ldots, N, . \tag{3.32}$$

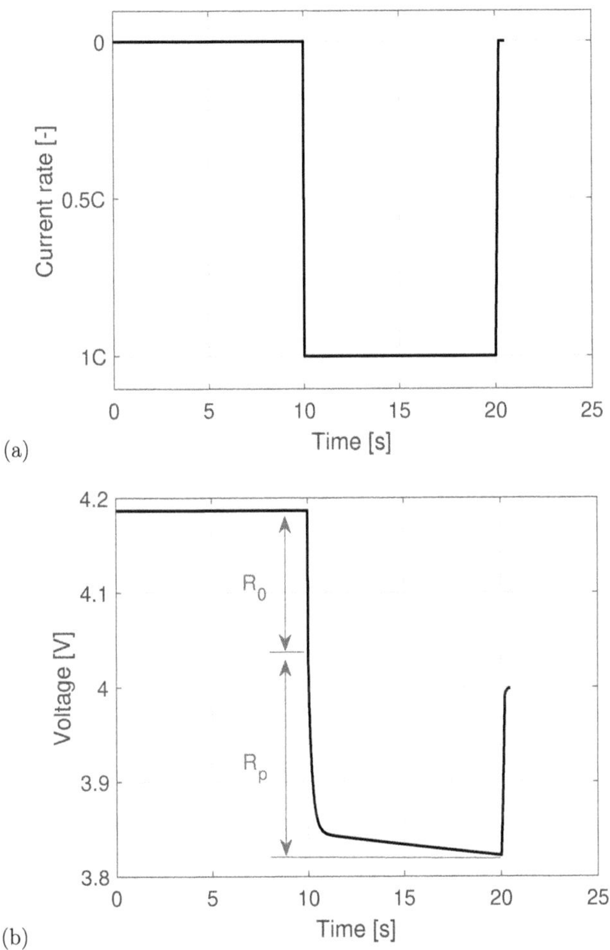

Figure 3.5: a) The current step for internal resistance measurement. b) A typical voltage response obtained across the cell.

The next step is to get the power loss $P_{l,j}$ using the *Joule* law

$$P_{l,j} = Ri_j \cdot i_j^2 \quad \text{for } j \in 1, \ldots, N. \tag{3.33}$$

To calculate the battery's energy loss, the ohmic resistance of the MOSFET and the contact resistance should also be taken into consideration. The energy loss is the integral of all power loses over time:

$$E_l(t) = \int_0^t \left(\sum_{j=1}^{N} Ri_j \cdot i_j^2(\tau) + \left(\sum_{j=1}^{N} R_{\text{DS,on}} + R_{\text{contact}} \right) \cdot i_{\text{bat}}^2(\tau) \right) \mathrm{d}\tau. \tag{3.34}$$

Substituting the value of i_j results in

$$E_l(t) = \int_0^t \left(\sum_{j=1}^{n_{\text{on}}} Ri_j + N \cdot R_{\text{DS,on}} + R_{\text{contact}} \right) \cdot i_{\text{bat}}^2(\tau) \, \mathrm{d}\tau. \tag{3.35}$$

Rearranging 3.35 and assuming that all battery cells have the same internal resistance, equation 3.35 is reformulated as follows

$$E_l(t) = \int_0^t (n_{\text{on}} \cdot Ri_j + N \cdot R_{\text{DS,on}} + R_{\text{contact}}) \cdot i_{\text{bat}}^2(\tau) \, \mathrm{d}\tau. \tag{3.36}$$

Chapter 4

Optimization theory

The term optimization generally refers to the search for an optimal solution to a problem. It involves calculating the best possible solution for a given problem by satisfying constraints expressing the limits on admissible solutions. The field of optimization is represented in many industries and economic domains. However, the suitable optimization method for a given problem heavily depends on its class and formulation. In general, a distinction is made between static and dynamic optimization problems, the difference of which is explained in this chapter.

4.1 Static optimization

The default formulation of a static constrained optimization problem is given as follows

$$\begin{aligned} \min_{\boldsymbol{x}\in\mathbb{R}^n} \quad & f(\boldsymbol{x}) \\ \text{s.t.} \quad g_i(\boldsymbol{x}) &= 0, \quad i=1,\ldots,p \\ h_i(\boldsymbol{x}) &\leq 0, \quad i=1,\ldots,q \end{aligned} \tag{4.1}$$

where $\boldsymbol{x}$ includes the variables that need to be optimized in order to minimize the function $f(\cdot)$ which represents the performance index. The functions g_i and h_i are equality and inequality constraints, respectively. They express the limit on the admissible solutions [57]. The *Lagrangian* function for this optimization problem is

$$L(\boldsymbol{x}, \boldsymbol{\lambda}, \boldsymbol{\mu}) = f(\boldsymbol{x}) + \sum_{i=1}^{p} \lambda_i g_i(\boldsymbol{x}) + \sum_{i=1}^{q} \mu_i h_i(\boldsymbol{x}), \tag{4.2}$$

with the Lagrange multipliers $\boldsymbol{\lambda} = [\lambda_1, \ldots, \lambda_p]^\top$ and $\boldsymbol{\mu} = [\mu_1, \ldots, \mu_q]^\top$.

To analytically solve this optimization problem, necessary and sufficient optimality conditions need to be determined. Using the Lagrange function defined in equation 4.2, the necessary optimality conditions for the problem in the form of (4.1) are given as follows

$$\begin{aligned} \nabla_{\boldsymbol{x}} L(\boldsymbol{x}^*, \boldsymbol{\lambda}^*, \boldsymbol{\mu}^*) = \nabla f(\boldsymbol{x}^*) + \sum_{i=1}^{p} \lambda_i^* \nabla g_i(\boldsymbol{x}^*) + \sum_{i=1}^{q} \mu_i^* \nabla h_i(\boldsymbol{x}^*) \\ h_i(\boldsymbol{x}^*) \leq 0, \quad i = 1, \ldots, q \\ g_i(\boldsymbol{x}^*) = 0, \quad i = 1, \ldots, p \\ \mu_i^* \geq 0, \quad i = 1, \ldots, q \\ \mu_i^* h_i(\boldsymbol{x}^*) = 0, \quad i = 1, \ldots, q \end{aligned} \tag{4.3}$$

Note that equation 4.3 is also known as *Karush–Kuhn–Tucker* (KKT) conditions. Whereas, the sufficient condition for an optimal solution is derived based on the analysis of the Hessian matrix of the Lagrange function $\nabla^2 L(\boldsymbol{x}^*, \boldsymbol{\lambda}^*, \boldsymbol{\mu}^*)$. Note that a point $\boldsymbol{x}^*$ that fulfills both conditions is a local minimum. However, in the case of a convex optimization problem, the calculation of the solution is simplified. In fact, a point $\boldsymbol{x}$ that satisfies the KKT optimality conditions represents not only a local minimum but even a global optimum. Thus, the second-order optimality condition evaluation is not necessary since it does not provide further information [57].

Since the stationary condition $\nabla_{\boldsymbol{x}} L(\boldsymbol{x}^*, \boldsymbol{\lambda}^*, \boldsymbol{\mu}^*) = 0$ can be evaluated analytically only in rare cases, numerical methods are generally used to solve the optimization problem. Various mathematical methods to solve such a problem are discussed in the literature. Depending on the optimization problem class, a suitable algorithm that solves the problem more efficiently should be used. Note that the class of the optimization problem is specified among others by the form of the objective and constraint functions. A widely discussed class of optimization problems is linear programming, which involves optimizing a linear objective subject to a finite number of linear constraints. The best-known numerical method for this special class is the *Simplex* algorithm. It uses the geometric properties from linear optimization problems, a polyhedron, with a finite set of extreme points (corner) as illustrated in figure 4.1.

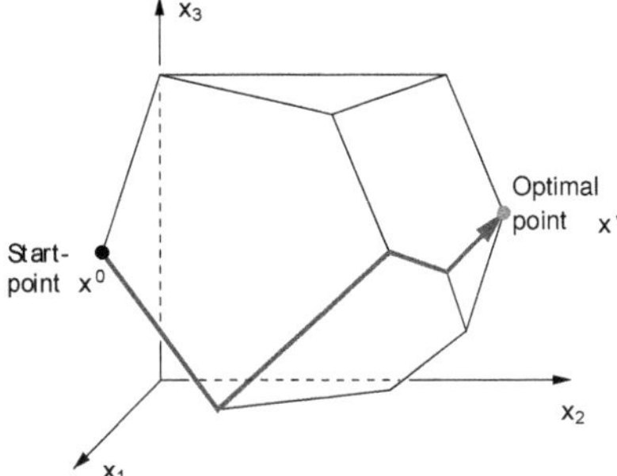

Figure 4.1: The Simplex algorithm starts at an initial vertex and moves along the edges of the polyhedron until it reaches the vertex of the optimal solution [57].

The optimal solution $\boldsymbol{x}^*$, if it exists, always lies on the edge of the polyhedron. If the $\boldsymbol{x}^*$ is unique, it lies on a vertex. The simplex method exploits this property by going from any corner of the permissible polyhedron along its edges to the optimum solution $\boldsymbol{x}^*$ without using the gradient information. Therefore, it is considered a non-gradient-based optimization method or direct search method.

Another category of numerical methods is the gradient-based optimization algorithms that use derivative information to determine a new point through an iteration rule, i.e., descent direction and step width. The iterations allow a sufficient decrease in the performance index function $f(\cdot)$ [58].

The purpose of a static optimization algorithm is to find at each point in time the point $\boldsymbol{x} \in \mathbb{R}^n$ that maximizes the instantaneous yield or minimizes the instantaneous cost by neglecting the impact of this decision on the global gain that can be reached.

Energy management strategies instead are often modeled by optimization problems that consider the time interval as a whole.

4.2 Dynamic optimization

Actually, a lot of economics and industrial systems require optimization over time. Such a system can be modeled using dynamic optimization theory, where a function of an independent variable that maximizes the gain (or minimizes the cost) should be found. In this section, the basics of dynamic optimization are discussed. Note that for the sake of simplicity, the minimization optimization problem will be tackled in this discussion.

A common approach to address dynamic optimization problems is the calculus of variations developed at the end of the 17th and the beginning of the 18th century. The most famous names associated with the calculus of variations are *Bernoulli*, *Euler* and *Lagrange* [57] [59]. Actually, the calculus of variations consists of controlling the state $\boldsymbol{x}$ of a system starting from an initial state $\boldsymbol{x}_0$ into a final state.

$$\int_{t_0}^{t_1} \phi\left(\boldsymbol{x}(t), \dot{\boldsymbol{x}}(t), t\right) \,\mathrm{d}t \tag{4.4}$$

is maximized over the time interval $[t_0, t_1]$.

The function $\phi\left(\boldsymbol{x}(t), \dot{\boldsymbol{x}}(t), t\right)$ represents the instantaneous cost at the time t. It depends on the state variable $\boldsymbol{x}$ and its rate of change $\dot{\boldsymbol{x}}(t)$. The goal is to minimize the aggregated instantaneous costs over the time interval $[t_0, t_1]$. The necessary optimality condition of the problem (4.4) is described by the partial differential equation of the form

$$\frac{\partial \phi}{\partial \boldsymbol{x}} - \frac{\mathrm{d}}{\mathrm{d}t}\left(\frac{\partial \phi}{\partial \dot{\boldsymbol{x}}}\right) = 0. \tag{4.5}$$

Equation 4.5 represents the *Euler-Lagrange* equation providing the first-order necessary condition for optimality. Each point that satisfies equation (4.5) and at the same time fulfills the constraint functions is a solution to the problem (4.4). Such a point is denoted by $\boldsymbol{x}^*(t)$.

The calculus of variations can be regarded as the forerunner of the maximum principle, which is a commonly used method in the control theory. Using the optimal control approach in the energy consumption issues is of particular interest when a trade-off between current and tomorrow's income is needed: a low instantaneous cost involves that only high costs are possible tomorrow and vice versa. In such a problem, the dynamic optimization algorithm searches for an optimal control path for the given control system that minimizes a cost functional over the optimization horizon. Basically, optimal control problems are formulated by a cost functional, a system dynamic equation, states and input constraints, and the initial and the final state conditions. It can mathematically be described as follows

$$
\begin{aligned}
\min_{\boldsymbol{u}(t)} \quad & J(\boldsymbol{x}(t), \boldsymbol{u}(t), t) = V(\boldsymbol{x}(t_\mathrm{f}), t_\mathrm{f}) + \int_{t_0}^{t_1} l(\boldsymbol{x}(t), \boldsymbol{u}(t), t)\mathrm{d}t \\
\text{s.t.} \quad & \dot{\boldsymbol{x}} = f(\boldsymbol{x}(t), \boldsymbol{u}(t), t), \qquad \boldsymbol{x}(t_0) = 0 \\
& g(\boldsymbol{x}(t_\mathrm{f}), \boldsymbol{u}(t_\mathrm{f})) = 0 \\
& h(\boldsymbol{x}(t), \boldsymbol{u}(t), t) \leq 0 \qquad \forall t \in [t_0, t_1]
\end{aligned}
\tag{4.6}
$$

where $J(\boldsymbol{x}, \boldsymbol{u}, (t)$ is the cost functional, which represents a criterion according to which the best control strategy is determined. The optimal solution for the problem (4.6) is thus the feasible control strategy that minimizes the cost functional. In fact, the cost functional integrates over time the running cost $l(\boldsymbol{x}(t), \boldsymbol{u}(t), t)$ and so it assigns total cost to each reachable state. The function $l(\boldsymbol{x}(t), \boldsymbol{u}(t), t)$ defines the cost that arises when applying the control variable $\boldsymbol{u}$ on the system dynamic equation. In contrast to the calculus of variations, the running cost function explicitly depends, among others, on the control variable $\boldsymbol{u}(t)$ and not on $x(t)$. Much more than in the calculus of variations, the cost functional could have a term that describes the final cost. This form of the cost functional is known as *Bolza*-form. In fact, the value of the control variable $\boldsymbol{u}(t)$ influences the system state $\boldsymbol{x}(t)$. Therefore, $\boldsymbol{u}(t)$ is the unknown variable to be optimized. The states shall include all system aspects relevant for the optimization so that knowing $\boldsymbol{x}(t)$ is not necessary to know how it is reached.

Compared to the calculus of variations, optimal control problems generalize the optimization problem by distinguishing between control variables and state variables [59]. In the calculation of variation, the state variables $\boldsymbol{x}(t)$ is optimally controlled at every point in time t by adjusting its change rate $\dot{\boldsymbol{x}}$. In the control theory instead, more general control variable $\boldsymbol{u}(t)$ is used to influence the change rate $\dot{\boldsymbol{x}}$. The relation between control variables and the change rate of system states is given by the system dynamic equation [60] as follows

$$\dot{\boldsymbol{x}} = f(\boldsymbol{x}(t), \boldsymbol{u}(t), t). \tag{4.7}$$

From equation 4.7, it is clear that an ordinary differential equation describes the dynamics of the state variable X.

In order to derive the necessary optimality conditions for the unconstrained optimal control problem, calculus of variations (equation 4.5) and *Lagrange* multiplier are used [59] [61]. These conditions can be written in compact form by the means of *Hamilton* function

$$H(\boldsymbol{x}, \boldsymbol{u}, \boldsymbol{\lambda}, t) = l(\boldsymbol{x}, \boldsymbol{u}, t) + \lambda^{\top} f(\boldsymbol{x}, \boldsymbol{u}, t). \tag{4.8}$$

The necessary conditions for the unrestricted optimal control problem (4.6) can be formulated using the Hamilton function as follows

$$\begin{aligned} \dot{\boldsymbol{x}} &= \frac{\partial H}{\partial \boldsymbol{\lambda}} = f(\boldsymbol{x}, \boldsymbol{u}, t) \\ \dot{\boldsymbol{\lambda}} &= \frac{\partial H}{\partial \boldsymbol{x}} \\ 0 &= \frac{\partial H}{\partial \boldsymbol{u}}, \end{aligned} \tag{4.9}$$

where $\boldsymbol{\lambda}(t)$ is the *Lagrange* multiplier.

The previous considerations assumed that the optimal control problem given by equation (4.6) is unlimited. In practice, however, restricted optimization problems are most common. In the following, the control variables shall satisfy control constraints in the form

$$\boldsymbol{u}(t) \in U \subset \mathbb{R}^q \qquad \forall t \in [t_0, t_1] \tag{4.10}$$

where U represents a subset of $\mathbb{R}^q$ that should be maintained at each time t. Restricted optimization problems can be taken into account by means of Pontryagin's maximum principle [57]. This principle is based on the Hamilton function and the canonical equations (4.9). However, the condition $\dfrac{\partial H}{\partial \boldsymbol{u}} = 0$ loses its validity as a necessary optimality condition and should be replaced by

$$H(\boldsymbol{x}^*, \boldsymbol{u}^*, \boldsymbol{\lambda}^*, t) = \min_{\boldsymbol{u}(t) \in U} H(\boldsymbol{x}^*, \boldsymbol{u}, \boldsymbol{\lambda}^*, t) \tag{4.11}$$

In other words, the optimal control should minimize the Hamilton function [52]. Figure 4.2 illustrates this necessary optimality condition.

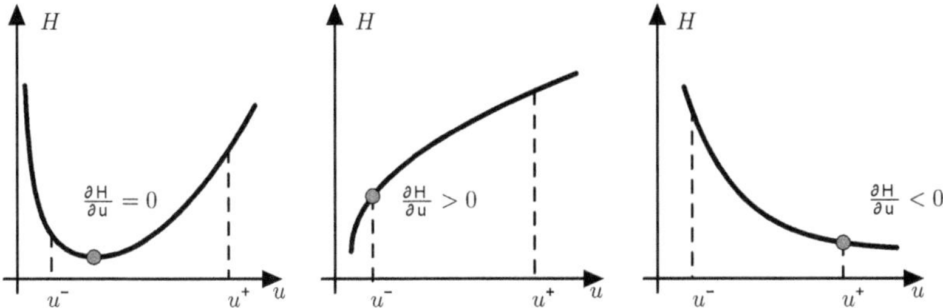

Figure 4.2: Illustration of the Pontryagin prinzip. The feasible control value that minimizes the Hamilton function is a solution to the optimal problem (marked in red) [57].

4.2.1 Numerical solution for dynamic optimization problems

A dynamic optimization problem can be generally described by equation (4.6). Although, no single approach is suitable for efficiently solving all the dynamic optimization problems due to their diversity. Therefore, different methods are reported to solve different types of optimal control problems. These approaches are divided into three classes: indirect methods, direct methods, and dynamic programming [62].

In the case of the indirect methods, the necessary conditions for optimality are first established. The problem is then formulated as a multi-point boundary value problem solved by collocation or multiple shooting methods. Hence, the optimization is done in an infinite-dimensional function space. The advantages lie in the accuracy of the solution found. However, the structure of the optimal control should be known in advance to handle the constraints. Furthermore, the obtained solution depends on the initial values of the problem parameters.

While indirect methods first optimize, then discretize the optimization problem, direct methods first discretize the problem then calculate the optimal solution [63]. As the name suggests, a direct method first transforms the optimization problem into a finite-dimensional nonlinear programming problem that can be solved afterward using the static optimization methods. In the case of a sequential procedure, the state, as a continuous variable, shall be determined in the prediction horizon using numerical integration methods. In the case of simultaneous procedures, the state variables are also discretized. Hence, the elements of the discrete state trajectory are also considered as optimization variables. The additional disassembly of the state trajectory creates a nonlinear programming problem with more decision-making variables. Handling constraints is much easier than by indirect methods. The main drawback of these methods is the accuracy of the results and the high computing effort required to run the optimization algorithm.

4.2.2 Dynamic programming

Dynamic programming is a well-known optimization method in the field of energy management strategies for electric vehicles. Such strategies are formulated as a discrete-time optimization problem

$$\begin{aligned} \min_{\boldsymbol{u}_k \in U_k} \quad & J(\boldsymbol{u}_k) = l_T(\boldsymbol{x}_T) + \sum_{k=0}^{T-1} l_k(\boldsymbol{x}_k, \boldsymbol{u}_k) \\ \text{s.t.} \quad & \boldsymbol{x}_{k+1} = f_k(\boldsymbol{x}_k, \boldsymbol{u}_k) \\ & k \in \{0, 1, \ldots, T-1\} \end{aligned} \tag{4.12}$$

where $\boldsymbol{x}_k$ and $\boldsymbol{u}_k$ are the system state and the control variable at stage k, respectively. The dynamic programming approach is based on the principle of optimality developed by Bellman for the numerical solution of optimal control problems:

An optimal policy has the property that, whatever the initial state and initial decision are, the remaining decisions must constitute an optimal policy with regard to the state resulting from the first decision. (See Bellman, 1957, Chap. III.3.). Figure

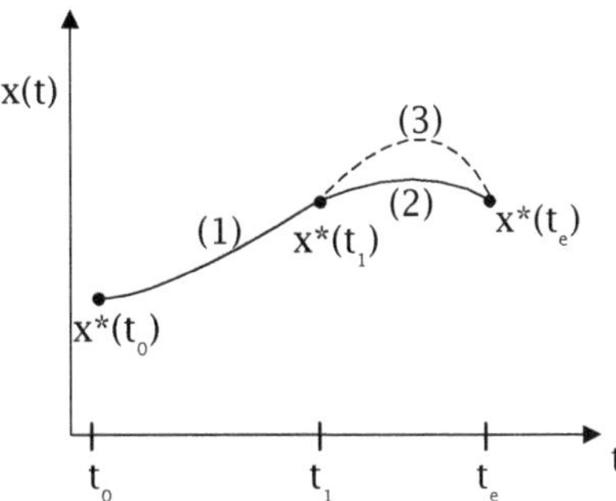

Figure 4.3: Illustration of the principle of optimality developed by Bellman: the sub-policy (2), which is a part of the entire optimal control policy going from $x^*(t_0)$ to $x^*(t_e)$, represents the optimal solution of the partial trajectory $[t_1, t_e]$.

4.3 gives a schematic representation of the Bellman principle. It is obvious that if the decisions (2) were not optimal for partial trajectory $[t_1, t_e]$, the corresponding cost is thus not optimal and could be further optimized by applying alternative decisions (3) yielding the optimal total cost. This leads to the fact that the entire trajectory from $x^*(t_0)$ to $x^*(t_e)$ including decisions (2) is not optimal, which contradicts the aforementioned assumption. As a result, a partial trajectory of an optimal trajectory is also optimal.

The Bellman principle reduces the dynamic optimization problem (4.12) to a sequence of static optimization problems which can be solved backward or forward in time.

The Dynamic programming method is well suited for mixed-integer optimal control problems that can be considered as a sequential decision process and satisfy *Markov* policy [61] [62]. It is when the current decision $\boldsymbol{u}$ can be made based on the current system state $\boldsymbol{x}$, independently of how $\boldsymbol{x}$ is reached. In other words, for a given state $\boldsymbol{x}_k$ reached at stage k, it does not matter for the future optimal decisions $\boldsymbol{u}_s$, $s \geq k$ how the state $\boldsymbol{x}_k$ is reached. This principle applies equally to time-discrete and time-continuous problems.

An important ingredient of the dynamic programming approach is the system state. It should include the system information needed to take the current decision regardless of the previous decisions and how the current state was reached. As the number of state variables strongly influences the dynamic programming algorithm's complexity, it is mandatory to keep this number small (up to 3) [61]. Otherwise, it is not possible to efficiently use dynamic programming due to the curse of dimensionality. Hence, the state is almost the important parameter of dynamic programming that should be carefully defined. The control variable $\boldsymbol{u}_k$ is an input parameter that enforces the system to move from state $\boldsymbol{x}_k$ to $\boldsymbol{x}_{k+1}$ by applying $\boldsymbol{u}_k$ to the system dynamic equation (4.12).

The approach of dynamic programming consists in structuring the control problem into multiple stages. Each one is solved using static optimization methods and by considering the results from the previous stage. Hence, the optimal control problem is not solved at once but the stages are solved sequentially one after the other. The optimization can be done backward or forward in time [64] [65]. A backward optimization process first involves evaluating the final stage and then solving the intermediate problems by moving back until the initial stage is reached so that all stages are considered. Thus, the backward dynamic programming algorithm can be described by the following recursive formula

$$\begin{aligned} V_T(\boldsymbol{x}_T) &= \quad l_T(\boldsymbol{x}_T) \\ V_k(\boldsymbol{x}_k) &= \quad \min_{\boldsymbol{u}\in U}\{l_k(\boldsymbol{x}_k, \boldsymbol{u}_k) + V_{k+1}(f_k(\boldsymbol{x}_k, \boldsymbol{u}_k))\} \\ k &\in \{0, 1, \dots, T-1\}. \end{aligned} \tag{4.13}$$

Also known as *Bellman* equation, the formula (4.13) includes the cost-to-go function that represents the optimal cost to move from the state $\boldsymbol{x}_k$ to the final stage. This function only depends on its previous value and the function $l_k(.)$ which describes the transition cost when moving from the state $\boldsymbol{x}_k$ to $\boldsymbol{x}(k+1)$. In the end, a forward recovery is needed that evaluates the trajectories calculated previously in order to reconstruct the optimal policy. This policy ensures the global optimum as it fulfills the *Bellman* equation representing the necessary and sufficient conditions for the global optimum.

An optimization problem in the form of (4.12) can also be solved forward in time, namely using the forward dynamic programming algorithm. Similar to backward dynamic programming, the optimization problem is divided into sub-problems. However, the forward algorithm starts the evaluation from the initial stage and moves one stage forward at a time until the final stage is reached. Thereby, a backward recovery is needed to determine the optimal policy. The recursions in equation 4.14 characterize the forward dynamic programming algorithm

$$\begin{aligned} V_0(\boldsymbol{x}_0) &= \quad V_{\text{init}} \\ V_{k+1}(f_k(\boldsymbol{x}_k, \boldsymbol{u}_k)) &= \quad \min_{\boldsymbol{u}\in U}\{l_k(\boldsymbol{x}_k, \boldsymbol{u}_k) + V_k(\boldsymbol{x}_k)\} \\ V_T(\boldsymbol{x}_T) &= \quad \min_{\boldsymbol{u}\in U}\{l_T(\boldsymbol{x}_T) + l_{T-1}(\boldsymbol{x}_{T-1}, \boldsymbol{u}_{T-1}) + V_{T-1}(\boldsymbol{x}_{T-1})\} \\ k &\in \{1, \dots, T-2\}. \end{aligned} \tag{4.14}$$

Generally, dynamic programming is applicable for deterministic as well as for stochastic optimal control problems. In certain cases, only one variant is feasible, e.g., stochastic problems can only be solved using backward dynamic programming. On the other hand, only the forward dynamic programming approach can be applied to optimization problems with an unknown final cost, namely, when the cost of a terminal state remains unknown until the entire control path calculation includes all stages. Forward dynamic programming starts from the initial state and searches forwards in time only over the feasible states that can be reached [66], which is not possible in backward dynamic programming.

Due to the global optimum and the lower computational required in most applications, dynamic programming is one of the most used numerical methods for optimizing operating strategies. Compared to the numerical methods presented above, dynamic programming has the advantage that a globally optimal solution can be provided, as the whole state space is searched. Another advantage of the dynamic programming method is that it can be applied to optimization problems with discrete states and discontinuous control variables. Hence, it is well suited for mixed-integer optimal control problems that can be considered as a sequential decision process.

Due to the curse of dimensionality, this powerful method only applicable to a problem with low state space dimension (less than 4) [6]. Besides dynamic programming, other numerical methods are used to optimize energy management for electric vehicles, such as genetic algorithms [67] [68].

Chapter 5

Noncausal strategy for global optimality

The static optimization approach is suitable for problems where today's output only affects today's profit. In this case, it is only necessary to look at its impact on today's profit and try to maximize it. However, when a decision made today impacts the cost or yield of tomorrow, then the control system should be formulated as a dynamic optimization problem. This approach is motivated by the fact that the current driving environment and the driver's driving style influence battery efficiency. Namely, the current peaks exhibited by a driving cycle cause a significant increase in power loss, affecting the battery efficiency. Therefore, the energy management strategy coordinates the cell modules in the reconfigurable battery by taking into consideration the dynamic of the power profile so that an optimum operation is achieved.

To this end, the strategy should provide a trade-off between energy loss and remaining energy so that the energy loss must be smaller than the energy profit obtained due to cell switching. In other words, the energy management strategy should find the optimal control policy for the cell switches that maximizes the amount of electrical energy transferred over a driving cycle. This strategy is developed based on mathematical models introduced in chapter 3 and using aspects from the optimization theory. As a result, battery efficiency can be enhanced, yielding the enhancement of the driving range and the energy throughput over the lifetime.

In this chapter, the control system will be first described as a multistage decision problem. Thereafter, an optimization algorithm is developed that calculates the global optimum to asses the benefits of dynamic cell switching. As this algorithm needs the prediction of power demand, the resulting strategy is a noncausal strategy. This strategy aims to maximize the total usable charge to enhance autonomy provided by the battery over a given driving cycle, rather than an instantaneous (at each instant) optimization of battery operation. This global optimum can also be used to benchmark causal strategies.

5.1 Formulation of the optimization problem

The energy management strategy developed in this chapter is a multistage decision process that can be described by a discrete-time optimal problem as follows

$$
\begin{aligned}
\min_{\boldsymbol{u}_k \in U_k} \quad & J(\boldsymbol{u}_k) = F(\boldsymbol{x}_T) + \sum_{k=0}^{T-1} l(\boldsymbol{x}_k, \boldsymbol{u}_k) & \\
\text{s.t.} \quad & \boldsymbol{x}_{k+1} = f(\boldsymbol{x}_k, \boldsymbol{u}_k), & \text{for } k \in \{0, 1, \dots, T-1\}, \\
& \boldsymbol{x}_0 = \boldsymbol{x}_{\text{init}}, & \\
& \boldsymbol{x}_k \in X_k, & \text{for } k \in \{0, 1, \dots, T\} \\
& \boldsymbol{u}_k \in U_k, & \text{for } k \in \{0, 1, \dots, T-1\}
\end{aligned}
\tag{5.1}
$$

In the following, a brief description of the most important ingredient of the optimal control (5.1) is reported.

Stage: The control strategy to be optimized is divided into stages, each of which includes a control decision that moves the system from a state to another. In fact, the operating time is sampled into periods $k = 0, \dots, T-1$, called stages. While $k = 0$ denotes the initial stage, $k = T-1$ is used for the final stage. The operating time is thus divided into T stages. In other words, T represents the length of the optimization horizon. At each stage, there is a number of alternatives (control variable). The selection of one feasible alternative represents stage decision. The presented energy management problem is deterministic as it is assumed that the power requests over the operating time are known accurately.

Control variable: In a multistage decision problem, one from many feasible alternatives in a stage will be selected for further calculation. This process is called stage decision. In a reconfigurable battery, the energy management strategy decides how many cells should be connected to the load at each stage. Hence, in the present optimization problem, the number of active cells is the control variable. The control variable values are thus a closed subset of $\mathbb{N}$. Constraints on the decisions should be considered. In fact, at each time, the battery current may not exceed the maximum current defined in the cell operating window. Furthermore, the minimum number of active cells depends upon the system rating voltage. The control variable is thus a natural number constrained by the minimum number of active cells

$$u_{\mathrm{min},k} = \max\left(n_{\mathrm{max,\ current}}, n_{\mathrm{min,\ voltage}}\right), \tag{5.2}$$

and the total number of battery cells N as an upper limit. While $n_{\mathrm{max,\ current}}$ is the minimum number of cells needed in order to keep battery current under the maximum allowed value, $n_{\mathrm{min,\ voltage}}$ represents the minimum battery voltage required. A subset of possible decisions U_k is associated with each stage and denotes the feasible controls available at stage k. Note that U_k varies with stage due to the variation of battery load and voltage of cells.

State: To solve the optimization problem using dynamic programming, the state shall include the system information needed to take the current decision regardless of the previous decisions. Hence, the knowledge of prior states becomes unnecessary. As implied in section 3.4, the equalization energy is the additional energy drawn from the best cells when bypassing the remaining battery cells. The balancing level of the battery is determined by dividing the amount of equalization energy reached at stage k by the one needed to balance the entire battery. This level summarizes the current status of the balancing process so that they fully specify the condition of decisions, i.e., at stage k a decision can be taken without the knowledge of how this balancing state was reached. Therefore, the balancing level represents the system state. Hence, the optimization problem presented in this work deals with one control law. Looking back at equation (3.28) the state transition rule can be described by the following discrete-time equation

$$x_{k+1} = x_k + \sum_{j=1}^{N} (1 - s_{j,k}) \cdot \Delta E_{j,k}, \tag{5.3}$$

where $\Delta E_{j,k}$ is the energy provided by the cell j at stage k and $s_{j,k}$ indicates the status of this cell, i.e., $s_{j,k} = 1$ if the cell j is active, and $s_{j,k} = 0$ is bypassed. This leads to the following

$$\sum_{j=1}^{N} s_{j,k} = u_k. \tag{5.4}$$

As the battery power is assumed to be constant over the sampling interval Δt_k, the variable $\Delta E_{j,k}$ can be obtained from

$$\Delta E_{j,k} = P_{j,k} \cdot \Delta t_k, \tag{5.5}$$

where $P_{j,k}$ is the power of the jth cell at stage k in case all battery cells would be active. Substituting the value of $\Delta E_{j,k}$ in the equation (5.3) results in

$$x_{k+1} = x_k + \sum_{j=1}^{N} (1 - s_{j,k}) \cdot P_{j,k} \cdot \Delta t_k. \tag{5.6}$$

The first term on the right in equation (5.6) represents the balancing rate. Let $g(u_k, P_{j,k})$ denote the balancing rate of the battery at stage k.

As in the control variable case, constraints defined in equation (5.2), boundary conditions, and state constraints should be defined for system states. In fact, for $n_{\text{on}}=n_{\text{Bal}}$, no more usable electrical energy remains in the battery, as shown in figure 5.1. That means each cell has reached 100% depth of discharge.

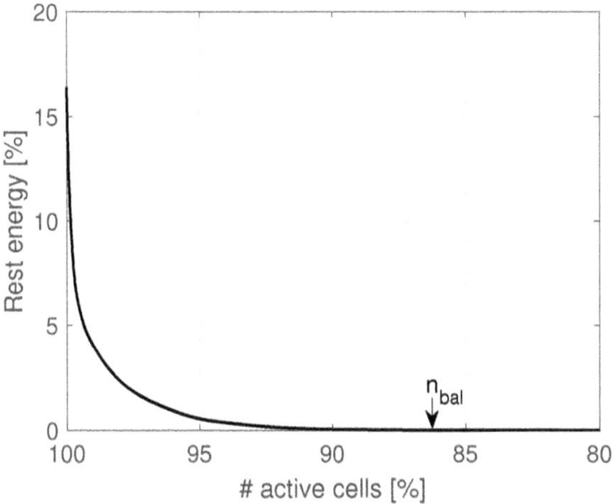

Figure 5.1: The remaining energy in the battery (normalized to the total energy stored in the battery) at the end of a complete discharge cycle as a function of the number of active cells (normalized to the total number of battery cells). The simulated battery is shown in 3.1.

As a result, a further decrease in n_{on} yields an increased energy loss without any further benefit. To avoid this effect, the balancing level should remain in a defined interval. Thus, the system state is restricted by zero downwards and x_{bal} upwards. In section 7.2, the maximum balancing level is explained in more detail. As explained in section 3.1, the balancing level influences the battery efficiency. Therefore, as battery efficiency should be maximized, the terminal state being free. As conclusion, the following constraints are valid for the system state

$$\begin{aligned} x_0 &= 0 \\ x_T &\leq x_{\text{bal}}. \end{aligned} \tag{5.7}$$

Cost functional: The cost functional in equation (5.1) is given in *Bolza* form. The first term represents the final cost, namely the cost associated with a state in the last stage T. The second term, which is the sum of function $l(\boldsymbol{x}(t), \boldsymbol{u}(t), t)$, defines the cost that arises when choosing the control variable $\boldsymbol{u}$. Note that the cost functional should be in the standard form shown in equation 5.1 to compute the optimal solution using dynamic programming. Hence, the cost functional should be formulated appropriately before developing the optimization algorithm. The energy management strategy developed in this chapter aims to maximize the amount of electrical energy transferred while discharging the battery. It is obvious that this energy, denoted by $E_{\text{out,elec}}$, is obtained when subtracting the accumulated energy loss from the energy extracted from the battery over the discharge period

$$E_{\text{out,elec}} = E_{\text{ext}} - E_{l,\text{bat}}. \tag{5.8}$$

The terms in equation (5.8) are formulated based on the model presented in chapter 3 and they can be defined thus far. The first term on the right denotes the accumulated energy loss over the discharge. Hence, it can be expressed as

$$E_{l,\text{bat}} = \sum_{k=0}^{T-1} l(x_k, u_k), \tag{5.9}$$

where $l(x_k, u_k)$ is the cost of the transition from stages k into the subsequent stage $k+1$ associated with the control u_k to system dynamic. A transition cost is the amount of electric energy transformed into heat over Δt_k. The amount of heat is given by *Joule* law as

$$l(u_k) = R_{\text{bat}}(u_k) \cdot i_{\text{bat}}(u_k, P_{\text{bat},k})^2 \cdot \Delta t_k. \tag{5.10}$$

where $P_{\text{bat},k}$ denotes the battery power over stage k, R_{bat} is the battery resistance, and i_{bat} is the battery current. The latter is described using the following mathematical formula

$$i_{\text{bat}}(u_k, P_{\text{bat},k}) = \frac{\sum_{j=1}^{u_k} V_{\text{oc},j} - \sqrt{\sum_{j=1}^{u_k} V_{\text{oc},j}^2 - 4 \cdot R_{\text{bat}}(u_k) \cdot P_{\text{bat},k}}}{2 \cdot R_{\text{bat}}(u_k)}. \tag{5.11}$$

Note that the battery current $i_{\text{bat}}(u_k, P_{\text{bat},k})$ and the transition cost $l(u_k)$ do not depend explicitly on the system state.

The first term in equation (5.8) corresponds to the final cost function $F(\cdot)$. This function defines a penalty term for each state reached in the last stage T. The idea is to add this term to the cost of each feasible terminal state. In the present case, the final cost function $F(\cdot)$ describes the relationship between a terminal state and the energy extracted from the battery. Hence, $F(\boldsymbol{x}_T)$ denotes the extracted energy associated with the terminal state $\boldsymbol{x}_T$.

Basically, the cost functional in discrete-time form sums up the running cost $l(\boldsymbol{x}(t), \boldsymbol{u}(t), t)$. Therefore, the value of the cost functional describes the total cost achieved. As the instantaneous energy loss is the running cost, the criterion according to which the computed control policy is evaluated is the accumulated energy loss during discharge. At this stage, the challenge is to maximize the amount of electrical energy transferred $E_{\text{out,elec}}$. Remember that maximizing x can be achieved through minimizing $-x$, and vice versa. Thus, the maximization problem is transformed into a minimization problem, namely, minimize $-E_{\text{out,elec}}$. As a result, the optimization problem examined in this work consists in finding a feasible control policy over all permissible controls that minimize the cost functional and satisfies the constraint as formulated bellow

$$\min_{u_k \in U_k} \quad J(x_k, u_k) = (-F(x_T)) + \sum_{k=0}^{T-1} R_{\text{bat}}(u_k) \cdot i^2_{\text{bat}}(u_k, P_{\text{bat},k}) \cdot \Delta t_k \tag{5.12a}$$

$$\text{s.t.} \quad x_{k+1} = f(x_k, u_k), \tag{5.12b}$$

$$x_0 = 0, \tag{5.12c}$$

$$u_k \in U_k, \tag{5.12d}$$

$$x_T \leq x_{\text{bal}}, \tag{5.12e}$$

$$k \in \{0, 1, \ldots, T-1\}. \tag{5.12f}$$

The energy management strategy is formulated as a convex constrained optimization problem. The transitions from one state to another be governed by $f(\cdot)$ defined in equation (5.6). The constraints on this optimization are that the decision at stage

k should belong to the set U_k of permissible decisions and the terminal state should not exceed x_{bal}.

5.2 Forward dynamic programming algorithm

In this section, an algorithm that efficiently solves the formulated optimization problem without any prior knowledge about the solution's structure is introduced. It is assumed that the power request over time is known prior to the optimization. As explained in chapter 4.2, dynamic programming is an approach to solve optimal control problems, which the presented optimization problem fits into. In fact, the aforementioned optimization problem is a multi-stage decision problem where the control variables are constrained to be positive integers. The system state is described by the recursive formula 5.6. In the case of a deterministic optimization problem, two different dynamic programming procedures can be applied to calculate the optimal policy. The most widely used procedure is the backward dynamic programming algorithm. Starting with the terminal cost, the backward dynamic programming recursively calculates the optimal cost-to-go for every feasible state at each stage k, for $k = T - 1, \dots, 0$. Instead, the forward dynamic programming consists in processing forward in time; from the initial stage $k = 0$ to the final stage $k = T - 1$, as shown in figure 5.2.

This approach is convenient for optimization problems with an unknown final cost, namely when the cost of a terminal state remains unknown until the entire control path calculation includes all stages. In the case of backward dynamic programming, the algorithm starts with the final stage in order to evaluate the optimal value function. As the final cost $F(\cdot)$ is only known when the feasible final state is calculated, the algorithm cannot finish the final state and sequentially including other stages backward in time. However, the forward dynamic programming uses the forward induction process by the evaluation of the optimal value function. Namely, the initial stage is the first stage to be solved based on the initial state condition.

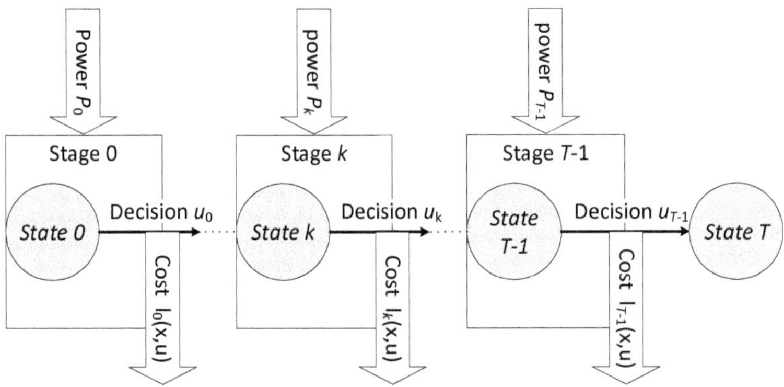

Figure 5.2: Illustration of the multistage decision process in the forward dynamic programming.

Using the system dynamic equation (5.12b), the algorithm moves forward one stage at a time until all stages are included. The feasible final states are thus determined, which allows the calculation of final costs. Therefore, the presented optimal control problem will be solved forward in time using the forward dynamic programming. Such an algorithm consists of a forward recursion and a backward recursion. These two phases will be dealt with in more detail below.

Forward recursion

The forward recursion process can be described by a network model shown in figure 5.3. The present optimization problem is thus to determine the shortest path through this network model. Control variables and states are represented by grids. U_k is the set of feasible decisions at stage k. Note that the dimension of decision sets varies with the power request.

The network model shown in figure 5.3 is an acyclic directed graph. It consists of nodes that represent system states. An edge indicates a state transition. Starting from the defined initial state x_0, the transition from a state x_k^i at stage k into the

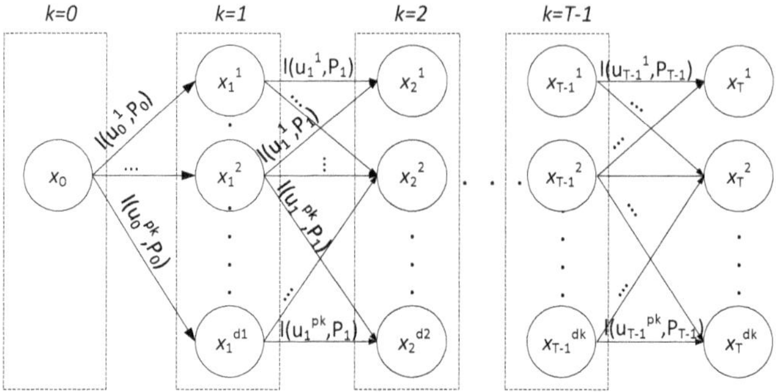

Figure 5.3: The proposed optimization problem is described as a network model.

state x_{k+1}^i from the next stage is done by applying a control $u_k \in U_k$ to the state transition equation (5.3). The edge weights reflect the cost's value at this stage given by $l(x_k, u_k)$. By checking the state restriction after each stage, the incremental expansion of the graph can be optimized.

As the proposed optimization problem is a one-dimensional dynamic system, two infeasible regions are to be considered. The first region includes states under the lower state limit. Because the objective is to equalize the battery cells during operation, the balancing level x_k should be higher or equal than the initial state x_0, $x_0 \leq x_k$, to be admissible. The lower limit of the second region of infeasible states is the maximum allowed equalization level x_{Bal}. This limit is derived using equation (3.18). In section 7.2, the maximum balancing level is explained in more detail.

Starting from the initial state $x_0 = 0$, the area of the reachable states is restricted due to the dynamic law (5.3). The state is a one-dimensional variable, so there are two limits for a feasible state. The upper limit is defined as the maximal reachable equalization level in stage k. This limit is derived using the following formula

$$x_{max,k} = x_{max,k-1} + g(u_{\min,k}, P_{j,k}). \tag{5.13}$$

The first term on the right represents the balancing rate which depends on the number of active cells and the power request. Note that both are kept constant over a stage k. The restricted system states area is illustrated in figure 5.4.

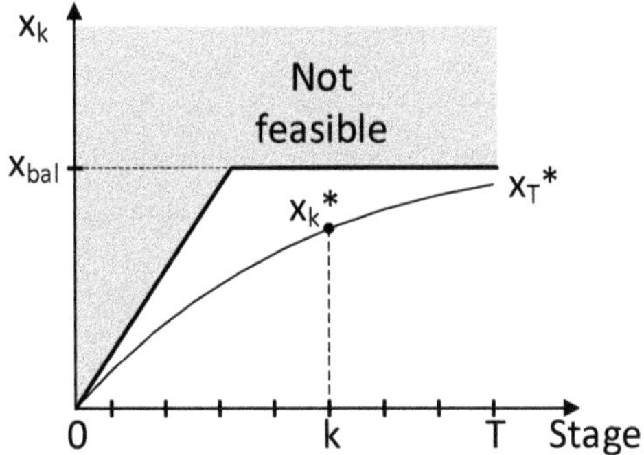

Figure 5.4: Illustration of the feasible area of the system states along the defined stages.

Decision process: In a forward recursion process, the first solved stage is the initial stage, and then all stages are included by moving forward one stage at a time. In fact, for each stage k and each state x_k^i, which represents the balancing level at stage k, the algorithm goes through the set of control variables U_k to find the number of active cells u_k that minimizes the total cost associated with the state x_k^i over the stages which are completed. The control variable is represented by a grid of size $p_k \times T$ where each column is incremented from the lower limit $n_{min,k}$, calculated in equation (5.2), to the upper limit N in single steps. Hence, the set of control variables U_k is defined as

$$U_k = \{u_k^1, u_k^2, \cdots, u_k^i, \cdots, u_k^{p_k}\}, \quad \text{with } k = 0, \cdots, T-1. \tag{5.14}$$

Note that the cost of a decision u_k^i is given by $l(x_k^j, u_k^i)$. Once the optimal value function was evaluated for all states of stage k, the system moves to the state x_{k+1}^i in the next stage $k+1$, which is calculated using the transition function (5.3). This state completely describes the system so that the next decision u_k^i can be made without knowing how this x_{k+1}^i state was reached.

State matrices: Analogously to control variables, the state space is incremented, forming a grid of size $d_k \times T$. Hence, the set of state variables X_k at each stage is defined as

$$X_k = \{x_k^1, x_k^2, \cdots, x_k^j, \cdots, x_k^{d_k}\}, \quad \text{with } k = 0, \cdots, T-1, \tag{5.15}$$

where x_k^j denotes the state variable in the discretized time-state space at the node with stage index k and state index i.

Calculation of the discrete states: As implied in chapter 4.2, the optimal control policy can be efficiently calculated when the optimization algorithm relies on dynamic programming techniques. This is because the optimization problem is broken down into subproblems sequentially resolved using different loops over control, state, and stage variables. Nevertheless, a further enhancement of the algorithm performance can be reached when these loops are replaced by matrix operations. Namely, the vectorization of the loops enhances the algorithm runtime, as shown in [69] [6].

In conventional dynamic programming, based on the system dynamic equation (5.12) and *Bellman* equation, three loops over stage, control- and state variables evaluate the optimal value function. In this work, the proposed algorithm uses matrices whose elements represent the repetitive operations of a loop. For instance, the control variable loop is replaced by the balancing rates vector $\boldsymbol{b} = (b^1 \; b^{p_k})^\intercal$. At each stage k, the vector elements b^i for $i \in \{1, \ldots, p_k\}$ are calculated as follows

$$b^i = \sum_{j=1}^{N} (1 - s^i_{j,k}) \cdot P_{j,k} \cdot \Delta t_k, \quad \text{where} \sum_{j=1}^{N} s^i_{j,k} = u^i_k, \text{ and } u^i_k \in U_k \tag{5.16}$$

Each element b^i represents the amount of balancing energy achieved over the stage k associated with the stage decision u^i_k. Note that u^i_k denotes the control variable in the discretized time-control space at the node with stage index k and control index i.

The calculation of state matrix is an iterative process. Using matrix operations, the set of admissible states X_k is calculated sequentially one after the other until the final stage is reached. In order to calculate the states of the stage $k+1$, the matrix $\boldsymbol{G}_k$ is defined as follows

$$\boldsymbol{G_k} = \begin{bmatrix} x^1_k & 1 & 0 & \cdots & 0 & 0 \\ x^1_k & 0 & 1 & \cdots & 0 & 0 \\ \vdots & \vdots & \vdots & \ddots & \vdots & \vdots \\ x^1_k & 0 & 0 & \cdots & 1 & 0 \\ x^1_k & 0 & 0 & \cdots & 0 & 1 \\ \vdots & \vdots & \vdots & \ddots & \vdots & \vdots \\ x^{d_k}_k & 1 & 0 & \cdots & 0 & 0 \\ x^{d_k}_k & 0 & 1 & \cdots & 0 & 0 \\ \vdots & \vdots & \vdots & \ddots & \vdots & \vdots \\ x^{d_k}_k & 0 & 0 & \cdots & 1 & 0 \\ x^{d_k}_k & 0 & 0 & \cdots & 0 & 1 \end{bmatrix}, \tag{5.17}$$

where x^j_k for $j \in \{1, \ldots, d_k\}$ are the elements of the state vector $\boldsymbol{X_k}$. The structure of the matrix $\boldsymbol{G}_k$ allows the calculation of all possible transitions between stage k and stage $k+1$ when multiplying $\boldsymbol{G}_k$ and $(1, \boldsymbol{b})^\intercal$. In fact, according to the process of matrix-vector multiplication, the dot product of vector $(1\ \boldsymbol{b})^\intercal$ with each of the rows of $\boldsymbol{G}_k$ is performed, as shown in equation (5.18)

$$\boldsymbol{G}_k \boldsymbol{A} = \begin{pmatrix} x_k^1 + 1 \cdot b^1 + 0 \cdot b^2 + \cdots + 0 \cdot b^{p_k} \\ \vdots \\ x_k^1 + 0 \cdot b^1 + 0 \cdot b^2 + \cdots + 1 \cdot b^{p_k} \\ \vdots \\ x_k^{d_k} + 0 \cdot b^1 + 0 \cdot b^2 + \cdots + 1 \cdot b^{p_k} \end{pmatrix}, \tag{5.18}$$

with

$$\boldsymbol{A} = \begin{pmatrix} 1 \\ \boldsymbol{b} \end{pmatrix}. \tag{5.19}$$

Let $\boldsymbol{Z}$ denote the output of the matrix-vector multiplication shown in equation (5.18). It is clear that each element of $\boldsymbol{Z}$ represents a transition from a state in stage k to the following state in the subsequent stage $k+1$. Once the matrix-vector multiplication occurs, the algorithm calculates the total cost of each element in $\boldsymbol{Z}$ by adding the running cost associated with the decision u_k^i to the optimal value of the start state x_k^i. Keeping in mind that it is not necessary to know how a state x_k^i it is reached, the state vector $\boldsymbol{X_{k+1}}$ is built by eliminating the repeated state elements. As a result, a grid of the feasible states is built, making the calculation process of the optimal control policy more efficient.

Parallel to the state vector setup $\boldsymbol{X_{k+1}}$, the algorithm defines the minimum cost associated with every calculated state at each stage. The interpretation of this minimum cost is slightly different between forward and backward dynamic programming. While the minimum cost represents the optimal cumulative cost when starting from the state x_k at the stage k and with $T-k$ stages to go until the terminal state is reached, the forward dynamic programming algorithm interprets the minimum cost associated with state x_k as the optimal cumulative cost when this state is reached starting from an initial state. This cost value associated with each state is calculated using the following recursive formula

$$V_{k+1}(x^j_{k+1}) = \min_{u^i_k \in U_k} \left(l(u^i_k, x^n_k) + V_k(x^n_k) \right) \quad \text{for} \begin{cases} x^n_k \in X_k \\ x^j_{k+1} \in X_{k+1} \end{cases} \tag{5.20}$$

with $x^j_{k+1} = f(u^i_k, x^n_k)$. The term in $\min(\cdot)$ function is nothing else but the total cost of each element in $\boldsymbol{Z}$ calculated previously. Also known as *Bellman* optimality equation for forward dynamic programming, the equation (5.20) is solved for each state that is an immediate successor of the initial state. The calculation proceeds in the forward direction until the entire state-stage space is covered. $V(\cdot)$ represents the minimum cumulative energy loss associated with each state denoted by cost-to-come reflects the minimum *Joule* losses generated when reaching that state from the initial state. These values are saved in a matrix called the cost-to-come matrix. A V^i_k element represents the minimum energy loss to move on the optimal trajectory from the state x_0 to x^i_k.

Note that no interpolation of the states is needed. In fact, the states in the state-stage grid are computed according to the system dynamic described in equation (5.6): $x_{k+1} = f(x_k, u_k)$. Consequently, starting from state x_k, which matches a grid point, the next state x_{k+1} corresponds exactly to one of the state grid points of stage $k+1$.

Optimal decision: Parallel to the calculation of cost-to-come, the algorithm searches over the set of admissible control values the decision that leads to the determined cost-to-come. This control value is saved in a policy function $\pi(\cdot)$ in order to identify the optimum control policy by a backward procedure. The total cost of each control policy $\pi = \{u_0 \dots u_k\}$ calculated in the forward induction is only known when calculating the energy return $F(x_T)$ of the final state matching this sequence of control actions which build an admissible policy.

An optimal decision map is built at the end of the forward recursion phase. This map is used in a backward recursion in order to determine the optimal control policy for the energy management strategy. The backward recursion is dealt with in more detail in the next chapter.

Optimal control policy

After calculating the space state and the cos-to-come matrix, the optimization algorithm begins searching for the optimal solution. To this end, the optimal states are identified by tracking back the paths computed, following the path of minimum energy loss. As mentioned in section 5.1 the present dynamic optimization problem has a free terminal state. Therefore, the terminal state will be handled as an additional optimization variable. This leads to the following

$$x_T^* = \arg \min_{x_T^j \in X_T} \left(-F(x_T^j) + V_{T-1}(x_k^j) \right) . \tag{5.21}$$

Note that $\left(-F(x_T^j) + V_{T-1}(x_k^j)\right) = V_T(x_k^j)$. After computing the optimal terminal state associated with maximum energy efficiency, a backward calculation procedure is started to determine the predecessor state in the optimal path. Starting at the optimal final state, the policy function $\pi(\cdot)$ identifies the optimal decision at stage $T-1$ that led to this state. Thereafter, the next state of the optimal path will be identified. The equation (5.22) represents the backward version of the system dynamic function defined in equation (5.6),

$$x_k = f^{-1}(x_{k+1}). \tag{5.22}$$

The predecessor optimal state in the previous stage is calculated by substituting the optimal state from the current stage into the backward transition function shown in equation (5.22). Once the optimal state from stage k is identified, the optimal control u_k^* entering this state is derived using the policy function again.

Using the backward transition function and the optimal policy function, the process computes the optimal stage decision sequentially until the initial state is reached. The optimal path calculation is then complete, and the optimal control policy for the dynamic optimization problem is determined. The admissible control policy which minimizes the cost function is denoted $\pi^* = \{u_0^* \dots u_k^*\}$.

Computational effort

The computational time of the optimization algorithm is a crucial feature. Therefore, vectors are used instead of *for*-loops. The vectorization of the dynamic programming algorithm reduces the time needed to calculate the optimal solution substantially [6]. The authors in [69] evaluated the performance of vectorized code and showed that such a technique could speed up an optimization algorithm with a large state grid by a factor of more than 800. The challenge of using vectors instead of *for*-loops is the high memory allocation required. To overcome this issue, only states that fulfill the boundary conditions (5.12) are included in the state vector x_k. The efficacy of the latter measure can be further enhanced when the stages are classified in descending order with respect to their energy content $E_k = P_{\text{bat},k} \cdot \Delta t_k$. Namely, starting with stages whose energy content is high allows the algorithm to detect and ignore paths leading to an unfeasible state early. Hence, the number of computations is further reduced.

Keeping in mind that $d_k \leq d$, where d is the maximal size of the state vector, the relationship between the size of a state vector and the number of stages used in the optimization should be investigated. Looking back at the equations (5.6) and (5.16) it can be seen that the difference between x^1_{k+1} and x^2_{k+1} resulting by applying u^1_k and $u^2_k = u^1_k + 1$ to the system at state x^i_k is given by

$$x^1_{k+1} - x^2_{k+1} = \sum_{j=1}^{N} (1 - s^1_{j,k}) \cdot P_{j,k} \cdot \Delta t_k - \sum_{j=1}^{N} (1 - s^2_{j,k}) \cdot P_{j,k} \cdot \Delta t_k. \tag{5.23}$$

After simplification

$$x^1_{k+1} - x^2_{k+1} = \sum_{j=1}^{N} s^2_{j,k} \cdot P_{j,k} \cdot \Delta t_k - \sum_{j=1}^{N} s^1_{j,k} \cdot P_{j,k} \cdot \Delta t_k. \tag{5.24}$$

Remember that $u^1_k = \sum_{j=1}^{N} s^1_{j,k}$, $u^2_k = \sum_{j=1}^{N} s^2_{j,k}$, and $u^2_k = u^1_k + 1$. The first term represents the energy that could be balanced over stage $k+1$ when u^1_k cells are active. The second term represents the energy that could be balanced over stage $k+1$ when

u_k^1+1 cells are active. Therefore, the difference between x_{k+1}^1 and x_{k+1}^2 represents the energy additionally balanced because that one more cell is bypassed. The smallest difference between two feasible states at stage $k+1$ is the electrical energy that the weakest cell would have delivered over Δt_k. As the state vector's size is finite, the smallest difference between two feasible states should be smaller or equal to the state vector's resolution. This restriction should be taken into consideration when building the multi-stage control problem. Therefore, in the first step, the driving cycle is transformed into a power histogram by counting the number of driving cycle points falling into the same power range, as shown in figure 5.5.

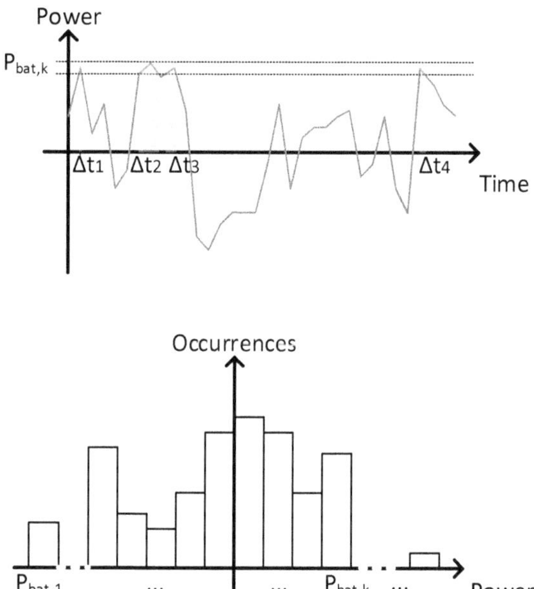

Figure 5.5: Transformation of a given driving cycle into a power histogram by counting the number of driving cycle points falling into the same power range.

In a second step, each power class that does not fulfill the aforementioned restriction is merged with the next one to form a power class. Note that the number of classes in the histogram represents the number of stages T of the optimization problem. Thus, T that corresponds now to the number of classes in the histogram, can be reduced without affecting the optimization result. Namely, the chronological order of power does not significantly impact the internal resistance and the open-circuit voltage off battery cells. The dependency of these cell parameters with the state of charge can be described by a step function [70]. Therefore, these parameters are assumed to be constant over one optimization horizon.

As a result, the complexity of the problem is reduced to $O(c) = \sum_{k=0}^{T-1} \cdot d_k \cdot p_k$ instead of $O(T \cdot d \cdot p)$ in the case of basic dynamic programming where the state and the control vectors have constant size over all stages [6]. Note that d denotes the state vector's size, and p is the number of possible control variables.

As conclusion, by implementing the measures explained above, the computational burden of the proposed algorithm can be reduced compared with the basic dynamic programming. The complexity can be further reduced when decreasing the number of stages T of the control problem. However, the quality of the obtained solution depends on the number of stage T and the state vector's size at each stage. Thus, choosing the state grid density and the number of stages is a trade-off between calculation time and accuracy.

Chapter 6

Application of noncausal strategy on the battery model

The proposed noncausal approach for optimizing the control strategy for a reconfigurable battery is validated in this chapter using simulations of realistic scenarios. Section 6.1 introduces the battery model used which corresponds to the reconfigurable architecture proposed in chapter 2. The battery model consists of 100 cells connected in series. The individual cells are simulated based on the dynamic cell model with three RC elements. The results of the implemented algorithm are shown in section 6.4. Section 6.5 is used to validate the noncausal strategy using simulations. A simulation study is conducted to validate the energy management strategy. The energy efficiency is compared for reconfigurable battery, conventional battery, and a battery whose cells are uniform. At the end of the chapter, the simulations' results are evaluated, and a conclusion is drawn.

6.1 Cell simulation model

In order to assess the performance of the energy management strategy proposed in chapter 5, an electrical model of the battery cell is needed. This model is built using the MATLAB simulation environment. In the following, the electrical model used to build the reconfigurable battery model will be presented along with the model validation results. By selecting the appropriate modeling approach, conflicting objectives such as accuracy, validity range, parameterization, and computing efforts should be considered. Different model types were discussed in chapter 2. As a compromise between model accuracy, parameterization, and computational effort, the equivalent circuit approach using linear and non-linear passive electric elements is the appropriate tool for electrical cell modeling to simulate the storage system's performance. The cell model determines all electrical cell variables needed as a function of the thermal conditions and the cell current. In addition to the terminal voltage V and the state of charge, the cell model provides the actual cell internal resistance Ri.

In the simulation study, an equivalent circuit model consists of an ideal voltage source, a series resistance, and three RC elements connected in series is used. The selected model is characterized by a simple parameterization and low computational effort with sufficient model accuracy.

Open-circuit voltage modeling: The open-circuit voltage depends not only on the temperature and state of charge but also on the previous load of the cell [71]. Thus, the cells exhibit an Open Circuit Voltage (OCV) hysteresis that depends on the current direction of the previous load and the relaxation time. The discharge characteristic indicates the open-circuit voltage for a considered state of charge, which was achieved by discharging after a full charge. The OCV curve depends on the previous cell operation. In fact, the open-circuit voltage after a previous charge is higher than the one after a discharge phase [71]. In this work, the OCV-value from the discharge characteristic is used as the battery's efficiency in discharge mode is investigated.

State of charge modeling: To determine the state of charge, we first need to integrate in time the current flows through the cell

$$\lambda(t_1) = \lambda(t_0) + \frac{\int_{t_0}^{t_1} i \, dt}{C}, \tag{6.1}$$

where $\lambda(t_0)$ is the initial state of charge and C is the actual cell capacity. This method is also known as *Coulomb Counting.* It provides an accurate state of charge values. In fact, in the simulation environment, the self-discharge is neglected and the integrated values correspond exactly the flowing current.

The inductance proposed by some research is not considered in the electrical cell model. In fact, an ideal inductive element has only an imaginary part and thus does not contribute to the power loss. Furthermore, the combination of series resistance in the range of mOhm and inductance in the range of microH requires dynamic high-frequency behavior with a low time constant. Therefore, the influence on the voltage response of the Lithium-ion cell can be neglected [72].

6.2 Model validation

The cell model presented in chapter 2 is implemented in the Simulink environment. Various measurements on a Lithium-ion cell are conducted in order to validate the equivalent circuit model used in this work. Other data than those considered in the parameterization are used for the validation. The cell data, e.g., terminal voltage, were collected every 100ms.

A dynamic load profile and different test temperatures are used to assess the model's accuracy. In fact, the cell was cycled under different ambient temperatures, including 5°C, 25°C, and 45°C and using Federal Test Procedure FTP-75 profile. This driving cycle characterizes urbane driving conditions so that the cell model is validated over a wide range of realistic operating conditions.

Figures 6.1 and 6.4 illustrate the simulated terminal voltage with experimental data obtained at different ambient temperatures. The solid line represents the measured terminal voltage, while the dotted line represents the model output voltage. The battery simulation results are compared afterward with real measurement data.

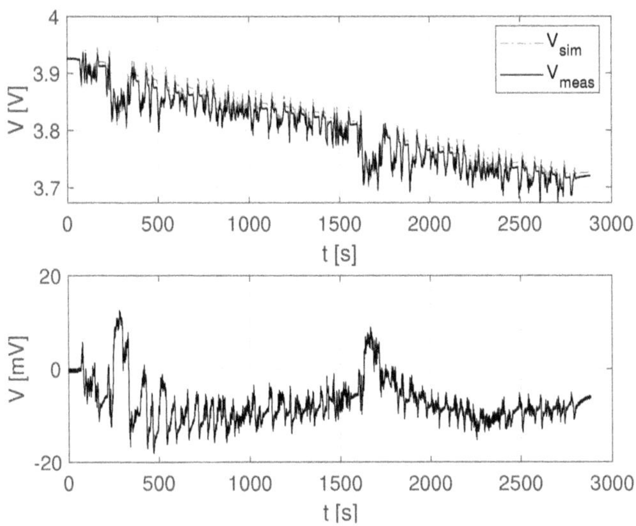

Figure 6.1: Comparison between experimental and modeling terminal voltage at 90% starting state of charge and 45°C ambient temperature.

Obviously, the cell model output agrees well with the measured terminal voltage most of the time and over different ambient temperatures. It is worth mentioning that the accuracy of the model may alter slightly depending on the cell temperature. The model response accuracy is further investigated by calculating the relative deviation and the root-mean-square error as follows

$$\mathrm{relErr} = \frac{V_{\mathrm{meas}} - V_{\mathrm{sim}}}{V_{\mathrm{meas}}}, \tag{6.2}$$

$$\mathrm{RMSE} = \sqrt{\frac{\sum_{i=1}^{n_{\mathrm{meas}}} (V_{\mathrm{meas},i} - V_{\mathrm{sim},i})^2}{n_{\mathrm{meas}}}}. \tag{6.3}$$

For temperatures above 25°C, simulation results match experimental data very

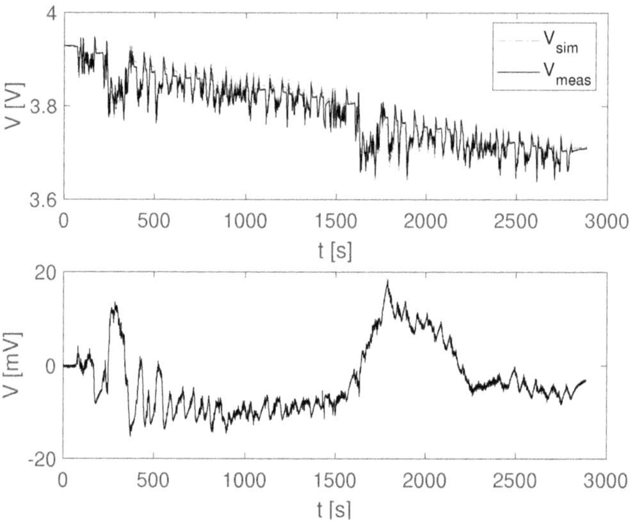

Figure 6.2: Comparison between experimental and modeling terminal voltage at 90% starting state of charge and 25°C ambient temperature.

well so that the root-mean-square error is under 13.2 mV. The difference between simulated and measured voltage is within +12 mV/ − 30 mV @45°C and +18 mV/ − 15 mV @25°C. These distribution ranges could also be read from the histogram of the voltage error, as shown in figure 6.7.

From figure 6.7, it is clear that the model has a lower accuracy at low temperature with a root mean square error of 46.7 mV. Nonetheless, the maximum relative difference between the simulated and the measured voltage remains below 2.8%, leading to an accuracy of more than 97%.

A summary of model accuracy is listed in tables 6.1 and 6.2.

The comparison results indicate that the model is suitable for validating the energy management strategies developed to enhance battery efficiency.

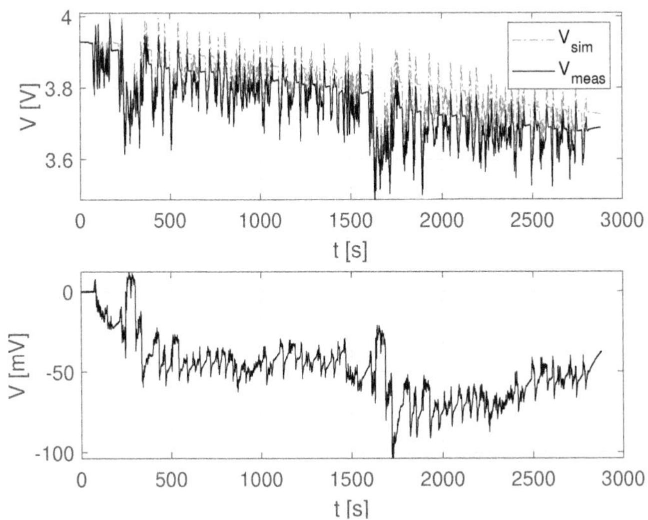

Figure 6.3: Comparison between experimental and modeling terminal voltage at 90% starting state of charge and 5°C ambient temperature.

Table 6.1: Model output error for different cell ambient temperatures: Absolute and relative error range.

Ambient Temperature [°C]	Absolute error range [V]	Relative error range [%]
45	$+0.012/-0.030$	$+0.325$ / -0.865
25	$+0.018/-0.015$	$+0.490/-0.404$
5	$+0.022/-0.103$	$+0.648/-2.797$

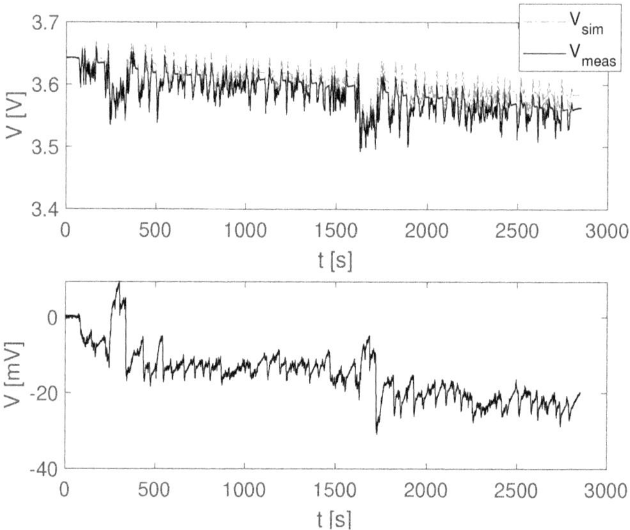

Figure 6.4: Comparison between experimental and modeling terminal voltage at 50% starting state of charge and 45°C ambient temperature.

Table 6.2: The root mean square error of the cell model for different cell ambient temperatures.

Ambient Temperature [°C]	RMSE [V]
45	0.0132
25	0.0076
5	0.0467

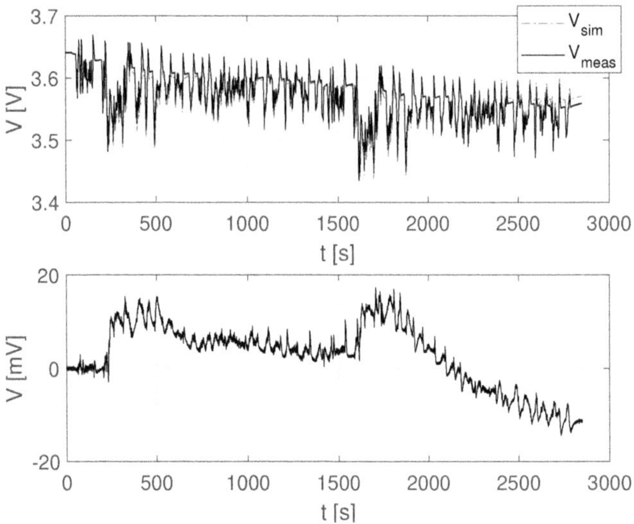

Figure 6.5: Comparison between experimental and modeling terminal voltage at 50% starting state of charge and 25°C ambient temperature.

6.3 Simulation setup

6.3.1 Battery cell capacities

As discussed in chapter 2, the Lithium-ion battery of an electric vehicle consists of a large number of cells connected in series. Some studies scale up the cell model to obtain a battery model. In such an approach, the cell-to-cell variation effects on battery performance are neglected. However, cell-to-cell variation is especially critical in a series configuration as the battery capacity is restricted by the weakest cell. In fact, when battery cells are inhomogeneous, they show different states of charge during operation. To model the effects of cell-to-cell variation on the battery efficiency, each cell in the battery model should have each own parameters, i.e., capacity and internal resistance. A lot of research is carried out in the field of cell-

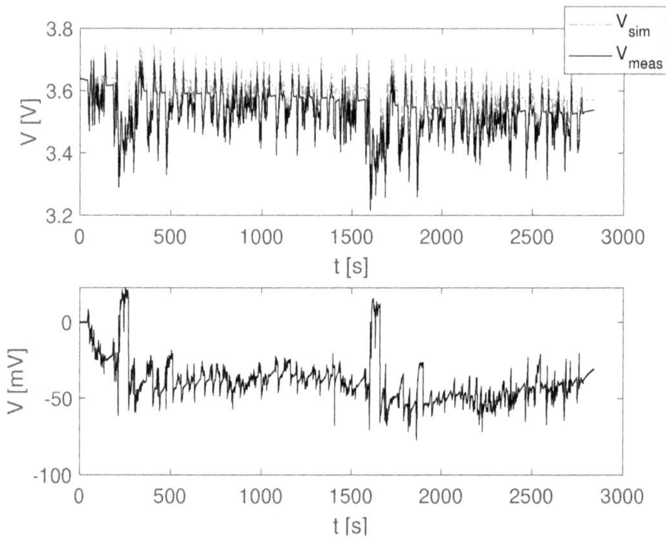

Figure 6.6: Comparison between experimental and modeling terminal voltage at 50% starting state of charge and 5°C ambient temperature.

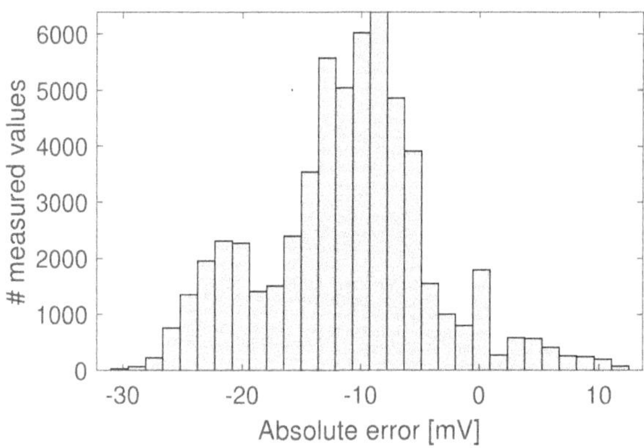

Figure 6.7: The histogram of the voltage error at 45°C ambient temperature.

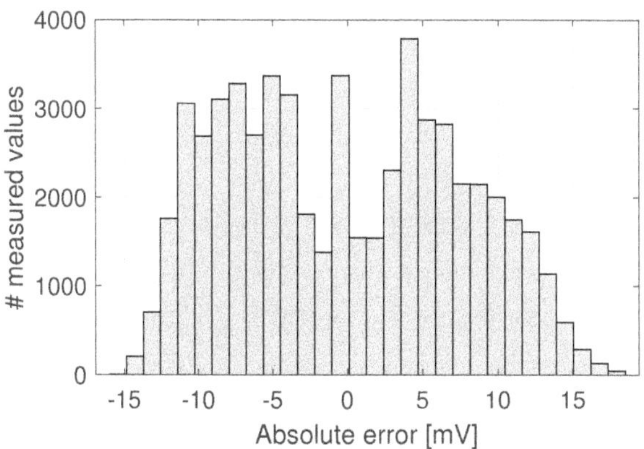

Figure 6.8: The histogram of the voltage error at 25°C ambient temperature.

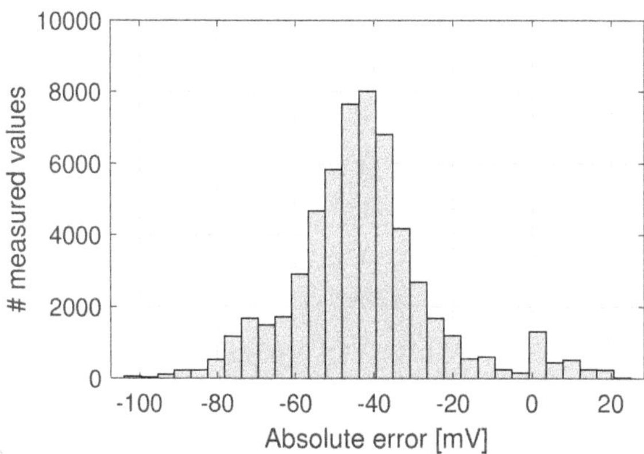

Figure 6.9: The histogram of the voltage error at 5°C ambient temperature.

to-cell variation. Due to the manufacturing process tolerances, the battery cells are not identical at the beginning of life. In addition, the cell-to-cell variation increases over time due to internal resistance disparity, temperature gradients, and operation conditions differences [73]. A detailed overview of the battery uniformity issues was provided in [2], including such topics as differences in self-discharging rate, aging behavior, and internal resistance.

The normal distribution is commonly used in the literature to describe the inhomogeneous cell parameters within the battery [73] [74] [75] [76]. The normal probability density function for the normal distribution of cell capacities is expressed as follows

$$p(C_i) = \frac{1}{\sigma_C \cdot \sqrt{2 \cdot \pi}} \cdot e^{\left(-\frac{C_i - \mu_C}{2 \cdot \sigma_C^2}\right)}, \tag{6.4}$$

with μ_C is the mean value of the capacity distribution, σ_C^2 is the variance, and σ_C represents the standard deviation.

In order to reduce the cell-to-cell variation, the cells are classified according to their initial capacities. The capacities and internal resistance of 20000 new battery cells having the same nominal capacity C_{nom} are measured in [74]. To measure the cell internal resistance, the current pulse method is applied. Cells were discharged with a current pulse having 2C amplitude and 5 s duration. The authors have shown that the distribution of the measured cell values fits the normal distribution, whereby the spread of the internal resistance is larger than the one of the cell capacity. While the capacity fits a normal distribution with a mean value $\mu_C = C_{\text{nom}}$ and a standard deviation of $\sigma_C = 1.3\%$, internal resistance distribution fits a normal distribution with a deviation of $\sigma_{Ri} = 5.8\%$ of the nominal value $\mu_{Ri} = R_{\text{nom}}$. Such a low capacity spread is only obtained when a costly classification process take place at the end of cell production. In fact, battery cells are classified according to their initial capacities to reduce the inhomogeneity within the multicell battery pack. Lack of classification influences the cell capacity distribution negatively. Namely, the capacities of unsorted cells still fit a normal distribution, but the standard deviation will be larger than the one from the classified cell. In [77] more than 700 new cells were measured. The authors have found that the capacity values are clustered into two clusters. Therefore, the distribution is fitted best by a bimodal distribution with a resulting standard deviation of 2.3%, which is smaller within a cluster, e.g., being by 1%.

In addition, if the battery is kept unused for a few weeks, the cells in the battery discharge due to self-discharge. Since the battery cells have different self-discharge rates, the drift of cells' states of charge becomes more significant, leading to the rise of cell-to-cell variation.

Besides the dispersion of the new cells' capacities and self-discharge rates, battery aging contributes to the rise of cell mismatch. Schuster et al. [75] investigated the effects of aging on the cell-to-cell variation based on the characterization of about 1500 cells. They proved the increase of the cell parameter disparity and the rise of outliers inside the battery with the aging progress.

6.3.2 Batteries data set

Simulating batteries with different spread values allows assessing the influence of cell mismatch on the energy management strategy's performance. Therefore, based on the investigations shown above, 5000 normal distributed capacities data sets with different coefficients of variation (CV) are randomly generated, as shown in figure 6.10. Note that the coefficient of variation has a uniform distribution and takes the following values $\{1\%, 2\%, \ldots, 10\%\}$. Furthermore, a capacity data set contains 100 values.

Analogously, 5000 normal distributed resistance data sets with different coefficients of variation (CV) are randomly generated, as shown in figure 6.10. Note that the coefficient of variation has a uniform distribution and takes the following values $\{1\%, 2\%, \ldots, 10\%\}$. Furthermore, a resistance data set contains 100 values. Since no correlation is known between the capacity distribution and the variation

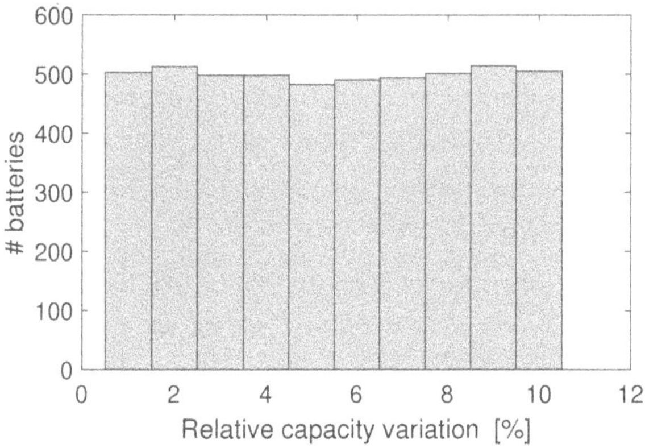

Figure 6.10: The Capacity standard deviation of the generated data sets.

of the internal resistance [73], cell resistance, and cell capacity distributions were considered to be independent.

It should be mentioned that the variability of cell parameters is expressed as a coefficient of variation in order to compare different data sets. Namely, the coefficient of variation defined as the ratio of the standard deviation to the mean

$$\mathrm{CV} = \frac{\sigma}{\mu} \cdot 100. \tag{6.5}$$

6.3.3 Driving cycles

Driving cycles are an important instrument in the validation of battery systems. On the vehicle level, a driving cycle represents the vehicle's speed over time. For a Lithium-ion battery, the driving cycle is given as a function of power. To obtain a power driving cycle, experimental tests could be needed [78]. A high sampling frequency ratio shall be used (e.g.,10Hz) to record the exact course of current when driving. In addition to the empiric approach, the authors in [79] suggested a model-based approach to obtain the battery power profile suitable for battery simulation. The introduced model calculates the battery power requirement from the velocity profile. This power profile represents the driving cycle.

The driving cycles considered in this study are the Federal Test Procedure FTP-75, which characterizes urbane driving conditions and an intern-developed driving cycle denoted by RDC, which describes rural and urban driving conditions. These driving cycles provide much better realistically battery test conditions than constant-current or constant-power cycling. Both driving cycles include load phases such as acceleration and phases in which energy is recovered by regenerative braking. Thus, charging and discharging alternate over time.

The set of driving cycles builds together with the generated data sets a simulation framework for the simulations that reflect reality and generates a large number of simulation results to validate the noncausal energy management strategy's performance.

6.4 Optimization algorithm results

6.4.1 Optimization model parameters in case of a relative capacity variation by 1%:

As explained in section 5.1, the cell equalization process is described by an optimal control problem. An optimization algorithm is developed and introduced in section 5.2. For a given driving cycle and set of cell parameters, i.e., capacity and internal resistance, the optimization problem calculates the optimal control policy that ensures these cells' discharge by maximum energy efficiency. As the energy efficiency analysis is used to benchmark the causal strategies, it is more convenient to limit the simulations to the discharge phase of the driving cycle.

The optimization algorithm searches the optimal control policy for a battery with 1% capacity variation when discharged using the FTP-75 driving cycle. Figure 6.11 shows the battery cell distribution and figure 6.12 illustrates the solution of the optimization algorithm.

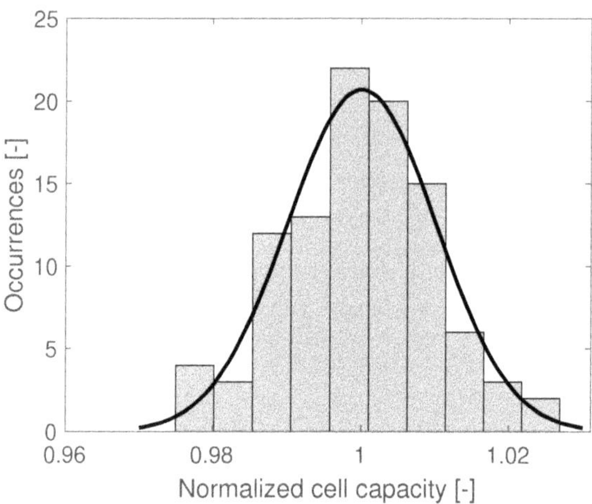

Figure 6.11: Normally distributed cell capacities are randomly generated (capacities are normalized to the mean capacity). The capacity variance is by 1%.

An analysis of the results yields two behaviors. The optimization algorithm attempts to reduce the energy loss by activating as many cells as possible during the high-power phase. Namely, the algorithm tries to augment the battery voltage and thus reduce battery current since the power loss is proportional to the current square. Along with this phase, the balancing rate is low as many cells are connected to the load. In contrast, when low power is required, the number of active cells is reduced to increase the balancing rate.

6.4.2 Optimization model parameters across different capacity variation values:

The same behavior is observed for different capacity spreads. The only difference is the number of cells that should be connected to the load. Namely, the more significant is the cell spread in the battery, the more cells are bypassed. Figure 6.13 visualizes this effect.

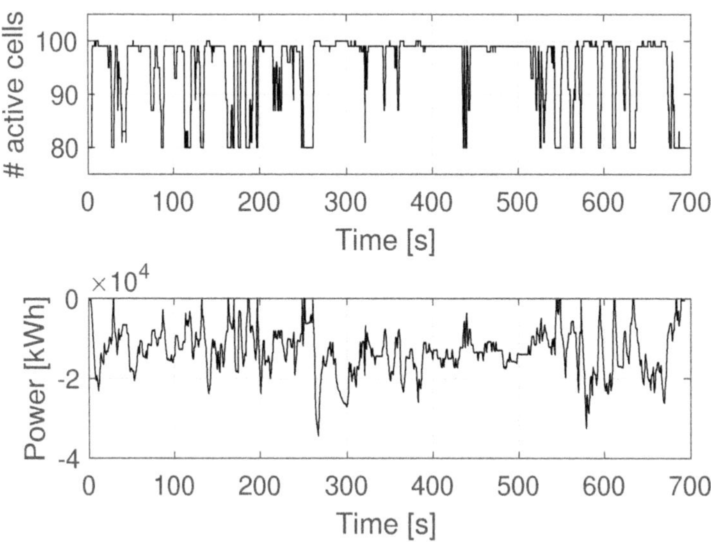

Figure 6.12: The optimal control policy (number of active cells as a function of time) used to discharge the battery shown in figure 6.11 using the discharge phase of the FTP-75 driving cycle.

As illustrated in figure 6.13, the number of active cells is kept at a minimum until power requirement about 11kW in the battery with 10% capacity variation. However, the number of active cells increases up power requirement of 7kW when the capacity spread is by 1%.

To get the average number of active cells over the driving cycle, optimal operational point values are added and divided by the number of the points. The next step is to build the average over the batteries having the same coefficient of capacity variation, as illustrated in figure 6.14.

The decrease in the average number of active cells is justified by the fact that the optimization algorithm attempts to bypass more cells in case of significant cell-to-

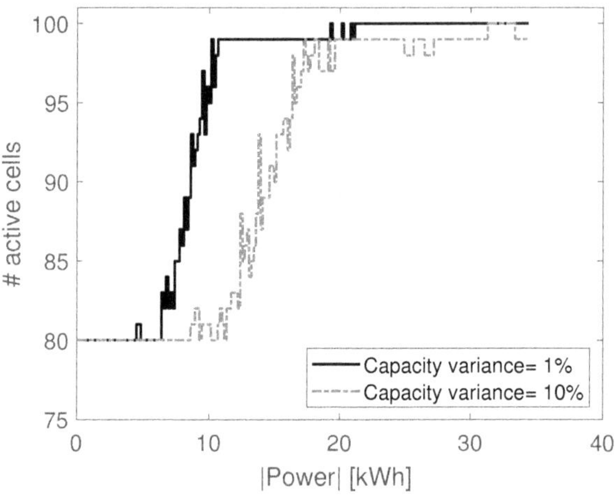

Figure 6.13: The optimal operational points of batteries with different capacity variance values: 1% and 10%. These results are obtained using the discharge phase of the FTP-75 driving cycle.

cell variation with respect to capacity. In fact, the larger the capacity spread within the battery, the more energy should be equalized. In order to achieve the balancing energy needed, the algorithm is obliged to bypass more cells compared to batteries with more homogeneous cells.

Another important aspect is the optimal terminal state. As implied in chapter 5, the purpose of the noncausal strategy is to provide a trade-off between energy loss and the remaining energy after reaching the cut-off criteria to maximize the battery's energy efficiency. A good way to visualize the trade-off aspect is to plot the optimal terminal state as a function of capacity variation, as shown in figure 6.15.

As illustrated in figure 6.14, the average number of active cells defined in the optimal control policy decreases when the capacity variation increase. As a matter of course, the optimal balancing energy also increases, as it can be seen in figure

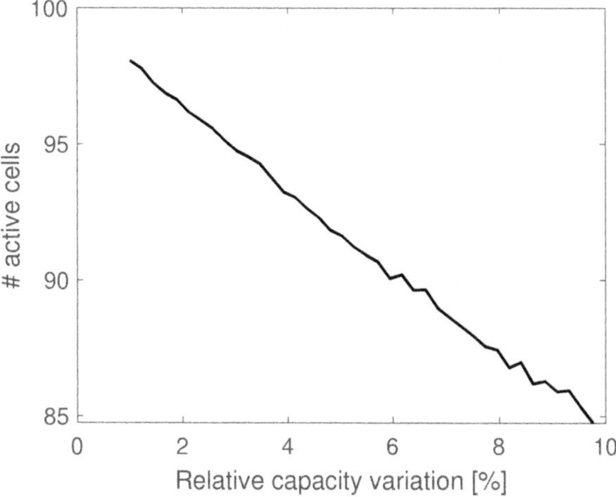

Figure 6.14: The mean value of the number of active cells as a function of the relative capacity variation when a battery is discharged using FTP-75 driving cycle. Mean value is the average over all simulated batteries falling in the same capacity variation class.

6.15. However, it is slower than the increase of the maximum balancing energy. In fact, the energy management strategy calculates the optimal number of active cells that maximizes the battery efficiency without fully discharging all battery cells, as in the case of high capacity variation.

6.5 Battery simulation results

This section is used to validate the developed noncausal strategy using simulations. While the cell model used in this work was experimentally validated, the investigation of the performance of the energy management strategy is mainly based on the simulation study.

The energy management strategy introduced in chapter 5 is evaluated using dis-

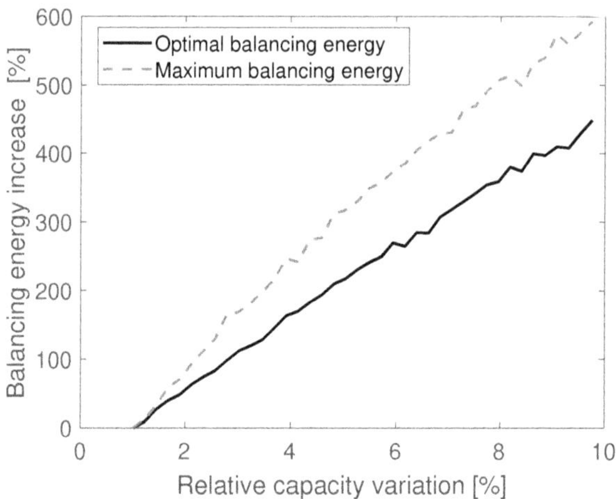

Figure 6.15: Illustration of the optimal balancing energy increase and the maximal balancing energy as a function of the relative capacity variation. The optimal balancing energy is the energy that should be balanced to maximize the battery's energy efficiency. The maximal balancing energy is the amount of energy that should be balanced to discharge every battery cell. These results are obtained when using the discharged phase of the FTP-75 driving cycle. The shown results are the average over all simulated batteries falling in the same capacity variation class.

charge simulations of batteries with different capacity spreads. As explained in chapter 3, it is meaningful to set the end of discharge criteria when the first cells reach the cut-off voltage. Therefore, the simulation is terminated as soon as the lowest cell state of charge falls below 1%. The energy efficiency is calculated afterward according to the formal (2.1). A simulation model for the conventional battery system with the passive balancing method was used to provide a baseline to assess the performance of the developed strategy. To ensure the same test conditions, both simulation models are actuated by the same cell parameters. Thereafter, the optimal control policy computed by the optimization algorithm is implemented in the reconfigurable battery model. Namely, the profile illustrated in figure 6.13 is used to construct a look-up table model integrated afterward in the battery model. Note that the implemented optimal control policy corresponds to the cell configuration. In the simulation environment, the requested battery current is calculated from the predefined power profile.

6.5.1 Comparison with the conventional storage system

After discharge simulations are conducted for each data set defined in section 6.3, the battery energy efficiency is calculated. Remember that the energy efficiency is the ratio of the electric energy transferred when discharging $E_{\text{out,elec}}$ to the total battery energy as shown in equation (2.1). While $E_{\text{out,elec}}$ is obtained as the integral of the product of battery terminal voltage and battery current from discharge start to cut-off

$$E_{\text{out,elec}} = \int_0^{t_e} V_{\text{bat}} \cdot i_{\text{bat}} \, \mathrm{d}t, \tag{6.6}$$

the authors in [36] assume that E_{tot} corresponds to the amount of energy that can be theoretically discharged and is described by the mathematical formula

$$E_{\text{tot}} = \int_{SOC_0}^{SOC_e} V_{\text{oc}} \cdot C \, \mathrm{d}SOC. \tag{6.7}$$

Figure 6.16 compares the energy efficiency of the reconfigurable battery and the conventional battery, which uses the passive balancing method.

It is clear that the energy efficiency of the reconfigurable battery increases by up to 2% compared to the conventional storage system when the cell spread remains under 1.3%. Note that this spread range corresponds to batteries at the beginning of life. The larger the capacity spread, the more significant the efficiency enhancement. In fact, the reconfigurable battery shows an efficiency of about 95% independently of the capacity spread. In contrast, the conventional battery's performance is affected when capacity spread increases so that its efficiency drops continually. For a capacity

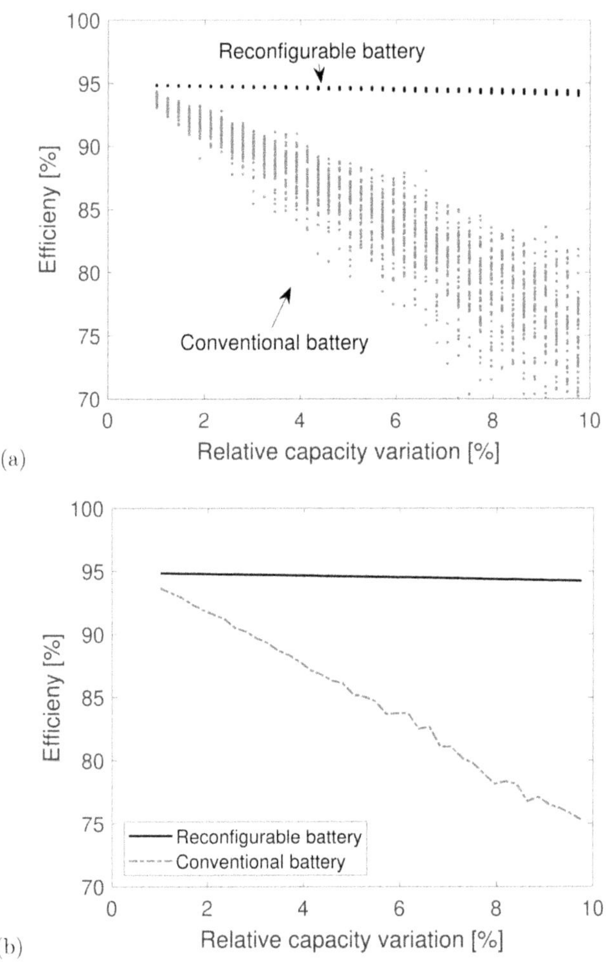

Figure 6.16: Comparison of the energy efficiency of the reconfigurable batteries and the conventional batteries during the discharge process using FTP-75 driving cycle. a) Representation of the efficiency values of the simulated batteries using dots. b) The mean value of the efficiency as a function of the relative capacity variation. A mean value is the average over all simulated batteries falling in the same capacity variation class is determined.

spread of about 8%, the efficiency of the conventional battery is by 78%, which means an efficiency enhancement of more than 15% when using the reconfigurable architecture along with the proposed energy management strategy. An important aspect can be seen in figure 6.16 a). Namely, the conventional battery's efficiency strongly depends on the distribution of capacities in the battery pack, i.e., within the same spread class, different efficiencies can be obtained. However, reconfigurable batteries falling in the same spread class show similar energy efficiencies. In fact, the efficiency variation remains in a narrow interval of 1%.

The efficiency of the reconfigurable battery is slightly decreased when cell spread increases. It is traced back to the increase of energy loss in the battery, as shown in figure 6.17.

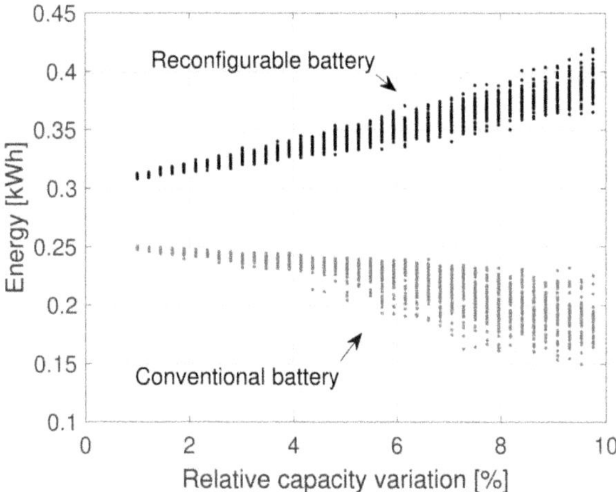

Figure 6.17: Comparison of the energy loss in the reconfigurable battery and the conventional battery when discharging both batteries using the RDC driving cycle.

A significant cell mismatch leads to a high capacity spread within the battery. The energy management strategy reduces the number of active cells in order to avoid weak cells reaching the cut-off voltage earlier. Consequently, battery terminal voltage decreases, leading to an increase in the battery current. As a result, the energy loss also increases. Figure 6.17 exhibits the energy lost in the form of heat for reconfigurable batteries with different capacity spread classes. Another aspect offered by the reconfigurable battery is extending discharge time compared to the conventional battery. In fact, the efficiency enhancement yields a significant improvement in the duration of battery operation. Figure 6.18 compares operating time periods of both battery systems.

It can be seen that the maximal operation time of the conventional battery drops when the capacity variance increases. However, the reconfigurable battery's operation time remains almost constant and is not affected by the increased capacity variance. Note that the simulated batteries have almost the same energy content but a different cell-to-cell variation.

6.5.2 Comparison with a battery having uniform cells

The calculated energy efficiency of the battery with uniform cells represents the benchmark for comparison on the performance of the noncausal energy management strategy. These simulation results reveal that the proposed optimization strategy enhances the battery efficiency and makes it close to one from the uniform battery pack, even by a significant capacity spread. The simulations demonstrate that the proposed energy management strategy offers the system robustness and allows operating along almost the ideal battery's energy efficiency. The marginal performance decrease of the proposed strategy is traced back to the increase in energy loss, as mentioned above (figure 6.17).

6.6 Conclusion

As the information about the future driving conditions must be available for the noncausal strategy, deterministic dynamic programming is limited to certain offline applications. Namely, this approach is usually used either as a benchmark for other

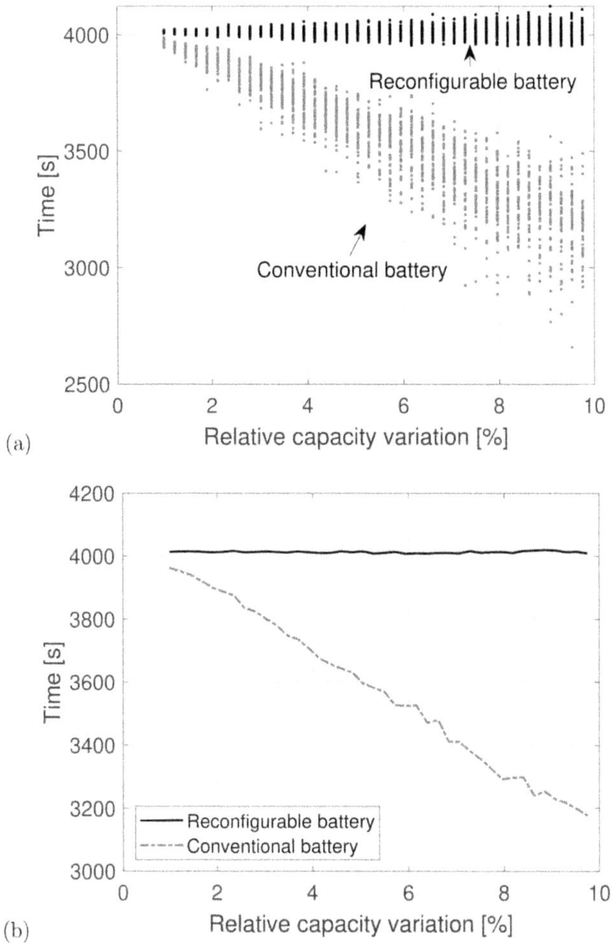

Figure 6.18: Comparison of the operating time of reconfigurable battery and conventional battery for the discharge phase of FTP-75 driving cycle. a) The operating time of each simulated battery. b) The mean value of the operating time as a function of relative capacity variation. A mean value is the average over all simulated batteries falling in the same capacity variation class is determined.

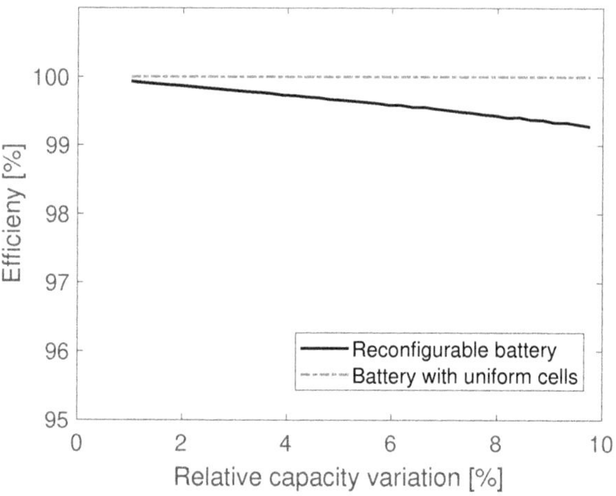

Figure 6.19: Energy efficiency comparison between the reconfigurable battery and the corresponding battery with uniform cells. The used driving cycle is the FTP-75 driving cycle. For better illustration, the energy efficiency of the reconfigurable battery is normalized by the one extracted from the ideal system.

operating strategies or to analyze the optimal operating mode to derive laws for rule-based operating strategies. In this context, causal approaches using the current driving conditions are introduced in the literature as suitable control methods for real-time applications.

Chapter 7

Causal optimal control strategies

In contrast to the dynamic programming approach presented in chapter 5, which considers the optimization problem as a whole, the optimization in this chapter is performed locally. The strategies introduced in this chapter are thus completely causal and real-time capable.

From open to closed-loop control: The optimization algorithm using the dynamic programming techniques needs information on the driving cycle. The optimal control optimization is like open-loop optimization, as shown in figure 7.1.

As mentioned previously, the decisions on the control variable u_k^* with $k = 0, \ldots, T-1$ are completely defined at the beginning of the battery operation based on the a priori given power profile. Therefore, if the power request during the operation deviates from which defined in the power profile, the battery efficiency could not be maximized. This is because this discrepancy in the power request was not taken into account. Consequently, in the case of an inaccurate power prediction in such an open-loop optimization strategy, the system states deviate from intended values expected by the optimization algorithm. Thus, the result is no more optimal. To overcome this issue, the optimal control value should be calculated at each sample time based on the current battery state and power requirement, namely, based on a closed-loop optimization, as shown in figure 7.2.

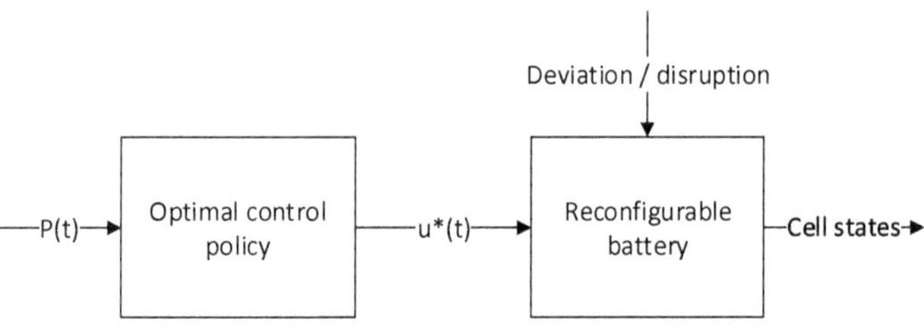

Figure 7.1: Open-loop control for the noncausal energy management strategy that maximizes the battery energy efficiency.

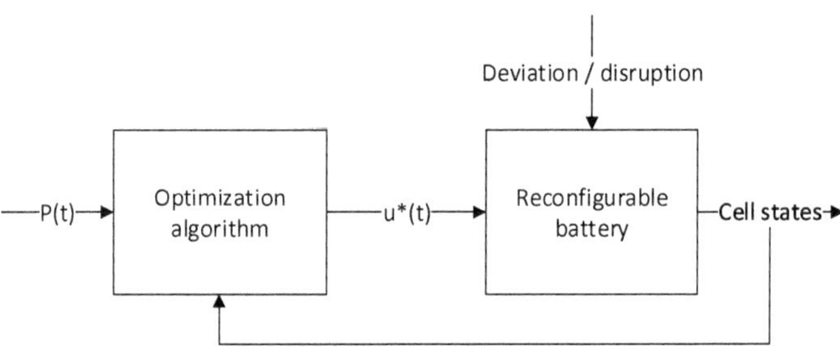

Figure 7.2: Closed-loop control for the causal energy management strategy that maximizes the battery energy efficiency.

In the course of this, two causal energy management strategies are developed as shown below in sections 7.1 and 7.2.

7.1 Maximize energy strategy (MES)

In this section, an energy management strategy that does not rely on the complete driving cycle's prediction is developed. The energy management system aims to maximize battery efficiency only based on the current driving conditions and the current system state.

Actually, simplifying the optimal control problem into a static optimization problem is discussed in the literature, usually noted as an equivalent consumption minimization strategy (ECMS). Paganelli et al [80] introduced an operating strategy for hybrid vehicles based on an instantaneous minimization problem that should be solved without any future information regarding the driving cycle. As no future information on the battery power requirement is available, the optimization algorithm calculates instantaneously the number of active cells that maximizes the operating time. Note that the number of active cells u and the power request P_{bat} are constant over the time step Δt. As a result, the following maximization problem is considered

$$\max_{u} \quad J(u) = \frac{F(u)}{P_l(u) + P_{\text{bat}}} \tag{7.1a}$$

$$\text{s.t.} \quad u \geq u_{\min}, \tag{7.1b}$$

$$u \leq N, \tag{7.1c}$$

$$u \in \mathbb{N} \tag{7.1d}$$

where u represents the number of active cells over the sample time Δt and $F(u)$ the energy that could be extracted from the battery when u cells would be connected to the load until the end of discharge. $F(u)$ is calculated based on the extracted energy model introduced in section 3.3.

Equation (7.1a) involves formulations that can not be analytically solved. Therefore, the proposed optimization algorithm enumerates at each time step all possible values of the cost functional $J(u)$ and then chooses the control value for which the cost functional is maximum. This method does not require any information about the gradient of the cost functional. As the algorithm evaluates all possible alternatives, the solution found represents a global optimum.

Real-time implementation: The algorithm calculates at each time the number of active cells n^*_{on} that maximizes the cost functional $J(u)$. It takes into account the currently requested battery power, and it does not need any future information on the driving cycle. Thereafter, the battery management system classifies the cells in decreasing order of their state of charge in order to connect the n^*_{on} best cells to the load over the sample time.

7.2 Optimize balancing strategy (OBS)

If the current battery power request information is not available for the energy management strategy, the energy management strategy needs to be further simplified. For this purpose, the control strategy is formulated as a static optimization problem where the function to be maximized does not depend on the battery power request. The static optimization problem has the following format

$$\begin{aligned} \max_{\boldsymbol{x}\in\mathbb{R}^n} \quad & f(\boldsymbol{x}) \\ \text{s.t.} \quad g_i(\boldsymbol{x}) &= 0, \quad i = 1, \ldots, p \\ h_i(\boldsymbol{x}) &\leq 0, \quad i = 1, \ldots, q \end{aligned} \tag{7.2}$$

7.2.1 Formulation of the optimization problem

The idea is to calculate the optimal number of active cells regardless of the battery power request. Therefore, a trade-off between rest energy and energy loss is no more feasible. Instead, the optimization problem consists of finding the number of active cells such that each cell in the reconfigurable battery reaches 100% depth of discharge and such that the amount of energy lost in the form of heat energy during battery operation is minimized. The heat dissipation is the power loss in the battery, which can be calculated using *Joule* Law

$$P_l(n_{\text{on}}, P_{\text{bat},k}) = R_{\text{bat}}(n_{\text{on}}) \cdot i_{\text{bat}}^2(u_k, P_{\text{bat},k}), \tag{7.3}$$

where $R_{\text{bat}}(n_{\text{on}}) = N \cdot R_{\text{ds,on}} + \sum_{j=1}^{n_{\text{on}}} R_j$ the total battery resistance.
Substituting the battery current i_{bat} given by the formula (5.11) into the equation (7.3) results in

$$P_l(n_{\text{on}}, P_{\text{bat},k}) = \frac{\left(V_{\text{oc,bat}} - \sqrt{V_{\text{oc,bat}}^2 - 4 \cdot R_{\text{bat}}(n_{\text{on}}) \cdot P_{\text{bat},k}}\right)^2}{4 \cdot R_{\text{bat}}(n_{\text{on}})}, \tag{7.4}$$

where $V_{\text{oc,bat}}$ is the battery open-circuit voltage obtained from

$$V_{\text{oc,bat}} = \sum_{j=1}^{n_{\text{on}}} V_{\text{oc},j} \tag{7.5}$$

Remember that the total battery resistance is composed of switches resistance, the sum of active cells internal resistance, and contact resistance

$$R_{\text{bat}} = N \cdot R_{\text{DS,on}} + \sum_{j=1}^{n_{\text{on}}} Ri_j + R_{\text{contact}}. \tag{7.6}$$

Using a *Taylor series* around $P_{\text{bat}} = 0$ the equation 7.4 is approximated [10]

$$P_l(u_k, P_{\text{bat},k}) = \frac{N \cdot R_{\text{DS,on}} + \sum_{j=1}^{n_{\text{on}}} Ri_j + R_{\text{contact}}}{\left(\sum_{j=1}^{u_k} V_{\text{oc},j}\right)^2} \cdot P_{\text{bat},k}^2. \tag{7.7}$$

Keeping in mind that the battery power request should be fulfilled, it is clear that $P_{\text{bat},k}$ can be omitted, and the number of active cells is thus the only control variable that affects the power loss. Namely, increasing the number of active cells leads to the reduction of power loss. This is because the term $\left(\sum_{j=1}^{u_k} V_{\text{oc},j}\right)^2$ in the denominator is proportional to the square of the number of active cells while the nominator term

$\left(N \cdot R_{\mathrm{DS,on}} + \sum_{j=1}^{n_{\mathrm{on}}} Ri_j + R_{\mathrm{contact}}\right)$ is proportional to the number of active cells. Therefore, minimizing the power loss is equivalent to maximizing the number of active cells. Consequently, the unconstrained form of (7.2) is expressed as

$$\max_{\boldsymbol{x}\in\mathbb{N}} \quad J(\boldsymbol{x}) = n_{\mathrm{on}}. \tag{7.8}$$

As mentioned above, each battery cell should reach 100% depth of discharge, so bounds have to be set on the battery charge throughput. Namely, when $Q_{\mathrm{th,bat}}(t_{\mathrm{e}}) \geq Q_{0,j}\ \forall\ j \in \{1,\ldots,N\}$, it can be guaranteed that each cell is fully discharged.

Looking back to the equation (3.15), the battery charge throughput is expressed as a function of the number of active cells. Therefore, the following equality constraint is defined

$$Q_{\mathrm{th,bat}}(t_e) = \frac{\sum_{j=1}^{N} Q_{\mathrm{th},j}(t_e)}{n_{\mathrm{on}}}. \tag{7.9}$$

Given this information, and noting that the number of active cells is constrained to be a non-negative integer, the optimization problem tackled in this section is a mixed-integer optimization problem which can be expressed as follows

$$\max_{\boldsymbol{x}} \quad f(\boldsymbol{x}) = x_1 \tag{7.10a}$$

$$\text{s.t.} \quad x_2 \geq Q_{0,\mathrm{m}}, \tag{7.10b}$$

$$x_2 = \frac{\sum_{j=1}^{N} Q_{\mathrm{th},j}(t_{\mathrm{e}})}{x_1}, \tag{7.10c}$$

$$x_1 \in \mathbb{N} \tag{7.10d}$$

with $Q_{0,\mathrm{m}} = \max_{j\in\mathbb{N}}\{Q_{0,1},\ldots,Q_{0,j},\ldots,Q_{0,N}\}$, x_1 denotes the number of active cells and x_2 denotes the battery charge throughput $Q_{\mathrm{th,bat}}(t_{\mathrm{e}})$ when the cut-off criteria is reached. Hence, $\boldsymbol{x}$ is given by

$$\boldsymbol{x} = \begin{pmatrix} n_{\mathrm{on}} \\ Q_{\mathrm{th,bat}(t_{\mathrm{e}})} \end{pmatrix} \tag{7.11}$$

7.2.2 Solution of the optimization problem

In order to analytically solve this optimization problem, the solution of the mixed-integer programming relaxation (7.10) should be first found. Namely, the linear problem relaxation arises by omitting the integer condition. In this case, the constraint (7.10d), which restricts x_1 to be a non-negative integer, is relaxed. This lead to the following static optimization problem

$$\max_{\boldsymbol{x} \in \mathbb{R}} \quad f(\boldsymbol{x}) = x_1 \tag{7.12a}$$

$$\text{s.t.} \quad x_2 \geq Q_{0,\mathrm{m}}, \tag{7.12b}$$

$$x_2 = \frac{\sum_{j=1}^{N} Q_{\mathrm{th},j}(t_\mathrm{e})}{x_1} \tag{7.12c}$$

As mentioned in chapter 4, *Karush–Kuhn–Tucker* (KKT) conditions represent the necessary optimality conditions for the aforementioned relaxed problem (7.12). First, the function f and the constraints are combined into the equation below

$$L(\boldsymbol{x}, \lambda, \mu) = x_1 + \lambda \cdot \left(x_2 - \frac{\sum_{j=1}^{N} Q_{\mathrm{th},j}(t_\mathrm{e})}{x_1} \right) - \mu \cdot (x_2 - Q_{0,\mathrm{m}}) \,. \tag{7.13}$$

Equation (7.13) is also known as the *Lagrangian* function. Taking the stationary condition results in

$$\nabla_{x_1} L(x_1^*, \lambda, \mu) = 1 + \lambda \left(\frac{\sum_{j=1}^{N} Q_{\mathrm{th},j}(t_\mathrm{e})}{(x_1^*)^2} \right) = 0, \tag{7.14}$$

$$\nabla_{x_2} L(x_2^*, \lambda, \mu) = \lambda x_2^* - \mu x_2^* = x_2 \cdot (\lambda - \mu) = 0. \tag{7.15}$$

The complementary slackness is as follows

$$\mu \cdot (x_2^* - Q_{0,\mathrm{m}}) = 0. \tag{7.16}$$

From equation (7.15) and the fact that $x - 2 \in \mathbb{R}$ it is clear that

$$\lambda = \mu. \tag{7.17}$$

Looking back at the equation (7.14), $\lambda \neq 0$ can be derived. Thus,

$$\mu \neq 0. \tag{7.18}$$

From equation (7.16) and equation (7.18), the optimal values that solve the problem are determined

$$x_2^* = Q_{0,\mathrm{m}}, \tag{7.19}$$

and

$$x_1^* = \frac{\sum_{j=1}^{N} Q_{\mathrm{th},j}(t_\mathrm{e})}{Q_{0,\mathrm{m}}}. \tag{7.20}$$

As the presented problem is a convex optimization problem, the point $\boldsymbol{x}^*$ that satisfies the KKT optimality conditions automatically represents a global maximum. Let n_{bal} denote the solution of the relaxed optimization problem (7.12). Hence, n_{bal} satisfies the constraint (7.12b) and maximizes the function $f(\cdot)$ shown in equation (7.12a). That means that connecting the n_{bal} best cells to the load allows to fully discharge all battery cells by minimizing the energy lost in the form of heat.

Rounding strategies

In order to solve the optimization problem 7.10, it is necessary to relax the integer constraint $x_1 \in \mathbb{N}$ to $x_1 \in \mathbb{R}$. The relaxed problem is solved afterward. In the best case, This solution can be a non-negative integer number which means that this value is a feasible solution for the mixed-integer problem (7.10). However, in most cases, the relaxed problem's solution is a real number that is not feasible for the mixed-integer problem. In this case, a rounding strategy is to be applied. The author in [81] introduced the branch and bound method to solve mixed-integer programming. The idea is to employ a graph consisting of branches and nodes as a framework for the solution search process. In contrast to the brute force enumerate method, the branch and bound method expands feasible and promising nodes only, stage by stage, in order to reduce the computational effort. The optimal solution has the better cost functional value while satisfying the constraints.

However, only x_1 is constrained to be a positive integer. Therefore, the presented approach consists of fixing this variable to the nearest integer less than or equal to the relaxed solution. The constraint on x_2 in equation (7.15) is thus satisfied.

Real-time implementation

The algorithm calculates the optimal number of active cells n_{on}^* instantaneously, using the optimization law given by equation (7.20) and rounding strategy. It takes into account the current state of charge of battery cells and does not need any information on battery power request. Figure 7.3 shows the corresponding closed-loop control.

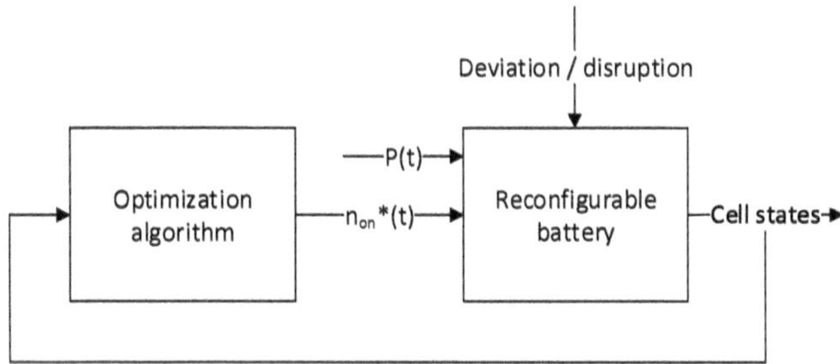

Figure 7.3: Closed-loop control for the causal energy management strategy that aims to optimize cell balancing.

The battery management system permanently classifies the cells in decreasing order of their state of charge in order to connect the n_{on}^* best cells to the load.

Chapter 8

Application of causal strategies on battery model

In this chapter, the energy management strategies developed based on the static optimization approach are validated by the noncausal strategy introduced in chapter 5. The battery model used in this simulation study is the same as the one from the noncausal strategy validation study. This model was introduced in section 6.1. It corresponds to the reconfigurable architecture proposed in chapter 2. Section 8.1 presents the results of both causal algorithms. Section 8.2 is used to validate these strategies using simulations. In order to make the simulation results of both noncausal and causal approaches comparable, the same data framework defined for the noncausal strategy simulation study is used in this chapter. In addition to the comparison between causal strategies and the passive balancing method, causal strategies are also compared with the *one-cell-off strategy* that represents the state of the art. At the end of this chapter, the simulations' results are evaluated, and a conclusion is drawn.

8.1 Results of the optimization algorithms

8.1.1 Results of the MES algorithm

In order to maximize optimal energy efficiency, the noncausal strategy connects more cells to the load during the high power phase than in the low power phase. The causal strategy cannot follow this logic as the power profile is unknown before the optimization is run. Instead, it calculates the optimal number of active cells based on the current power demand only. Figure 8.1 compares the solutions of both causal strategy and noncausal strategy.

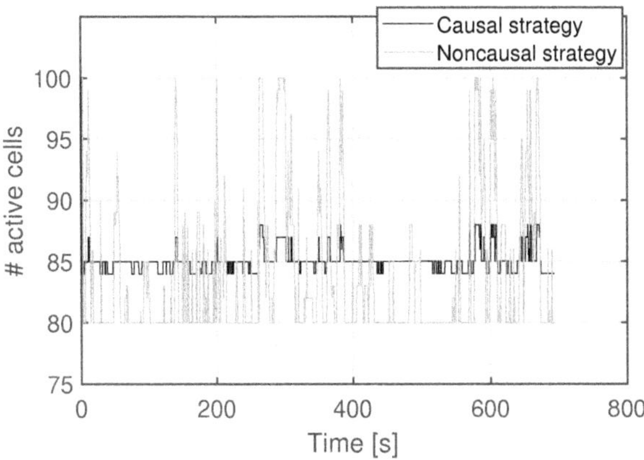

Figure 8.1: Comparison of the number of active cells calculated by the noncausal strategy and the causal strategy. The simulated battery is shown in figure 3.1.

As shown in figure 8.1, the causal strategy seems to be conservative compared with the noncausal approach. In fact, while the latter uses the maximum and the minimum number of active cells allowed to maximize the profit, the causal strategy fluctuates around a constant value due to the lack of a prediction horizon.

8.1.2 Results of the OBS algorithm

This strategy, which aims to optimize cell balancing, calculates the number of active cells for which each cell in the reconfigurable battery reaches 100% depth of discharge, and at the same time, the amount of energy lost in the form of heat during battery operation is minimized. This strategy calculates the optimal number of active cells using the equation (7.20) and the rounding strategy.

Figure 8.2 illustrates the output of the OBS for a battery with 5.3% capacity variation. Because OBS does not consider any information on battery power request

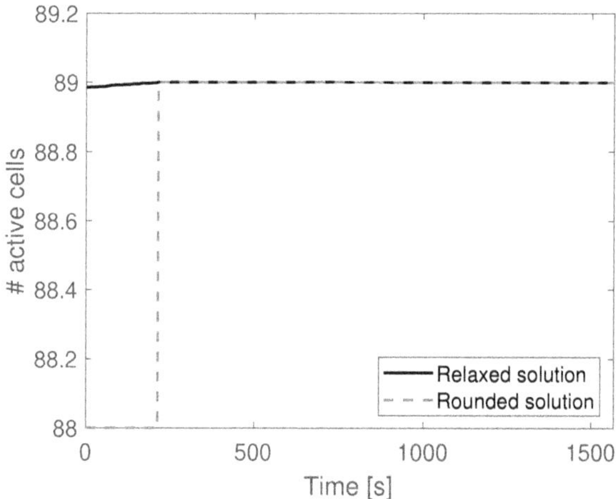

Figure 8.2: The optimal number of active cells calculated by the OBS and the corresponding rounded solution. The used power profile represents the discharge phase of the FTP-75 driving cycle. The simulated battery is shown in figure 3.1.

when computing the optimal number of active cells, the latter is almost constant and changes slightly during the operation. The rounding strategy fixes the output of the relaxed optimization problem to a positive integer. In fact, the number of cells was rounded to the nearest integer less than or equal to that value. This rounding yields a slight drift of the balancing state over time. Due to the closed-loop optimization, the OBS updates the optimal number of active cells. As expected, the optimal number of cells increases after a certain time.

In order to investigate the effect of the capacity spread increase on the optimal solution n_{bal}, the solution of the relaxed optimization problem is considered. First, n_{bal} is calculated for each battery from the simulation framework, as shown in figure 8.3. Thereafter, the average $\overline{n_{\text{bal}}}$ is built over the batteries having the same capacity

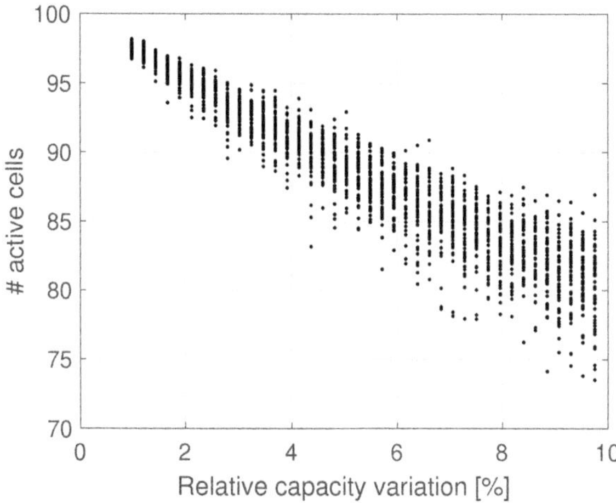

Figure 8.3: The relaxed number of active cells defined by the relaxed optimization problem for batteries with different capacity variations.

spread. Figure 8.4 shows the evolution of the mean value of balancing number for batteries with different capacity variation values.

It is clear that n_{bal} drops as the spread increases. This is because the worse the capacity spread within the battery, the more energy should be balanced. In order to achieve the balancing energy needed, more cells should be bypassed compared to batteries with better cell spread.

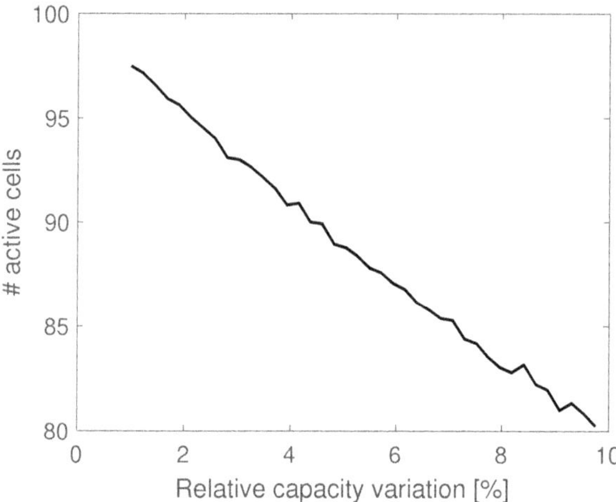

Figure 8.4: The mean value of balancing number as a function of capacity variation

8.2 Results of the battery simulations

Analogously to the noncausal strategy, the investigation of the causal strategies' performance is mainly based on the simulation study. The strategies developed in chapter 7 are evaluated through discharge simulations of the generated batteries. According to equation (2.1), the energy efficiency is calculated afterward. A simulation model for the conventional battery system with the passive balancing method was used to provide a baseline to assess the developed strategy's performance. To ensure the same test conditions, both reconfigurable and conventional battery models are actuated by the same parameters and the same inputs, such as cell parameters.

8.2.1 Efficiency comparison with conventional storage system

Figure 8.5 compares the energy efficiency of the reconfigurable battery implementing the developed noncausal strategy with conventional battery using passive balancing.

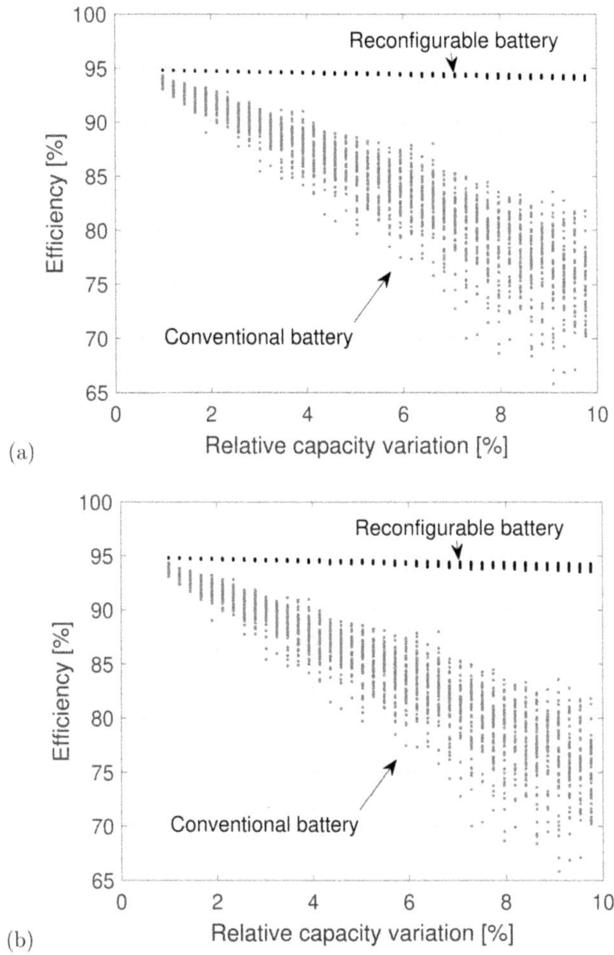

Figure 8.5: Energy efficiency comparison between the fixed cell battery and the reconfigurable battery with different strategies: a) MES, b) OBS. The used power profile represents the discharge phase of the FTP-75 driving cycle.

A similar effect as by the noncausal strategy is observed here. The worse the capacity spread, the more significant is the efficiency enhancement. Namely, reconfigurable battery shows an efficiency above 94.5% for capacity spread values up to 10%. In contrast, the conventional battery performance is affected by the rising spread so that the energy efficiency drops continually. Consequently, the presented causal strategies allow using more than 15% more energy than the conventional battery, including passive balancing.

8.2.2 Comparison of energy efficiency of causal strategies

As mentioned in chapter 5, the global optimum found by the optimal control can be used to benchmark the causal strategies. Figure 8.6 compares the energy efficiency of both noncausal and causal strategies, aiming to maximize the electrical energy transferred into the load.

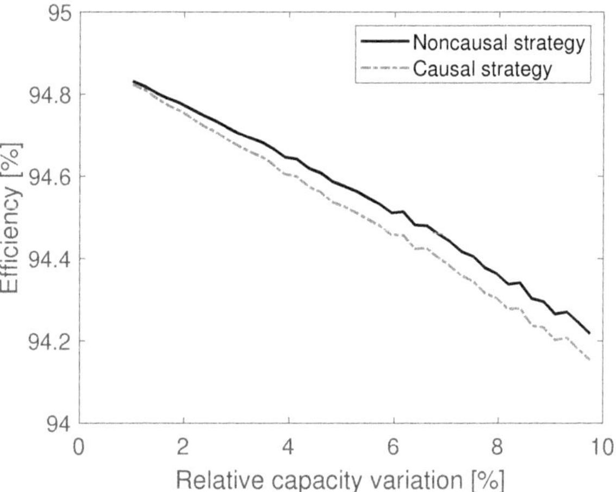

Figure 8.6: Energy efficiency of the reconfigurable battery with different energy management strategies. The used power profile represents the discharge phase of the FTP-75 driving cycle.

As it can be seen in figure 8.6, both results are close to each other. In the beginning, the energy efficiency values achieved by the different strategies are almost the same. The more significant the relative capacity variance, the better is the result of the noncausal strategy compared with the one from the causal strategy. Note that the difference between both strategies is minimal as it remains under 0.1%.

In the next step, the energy efficiency of both causal strategies is compared, as shown in figure 8.7. Both solutions are quite equivalent regarding energy efficiency.

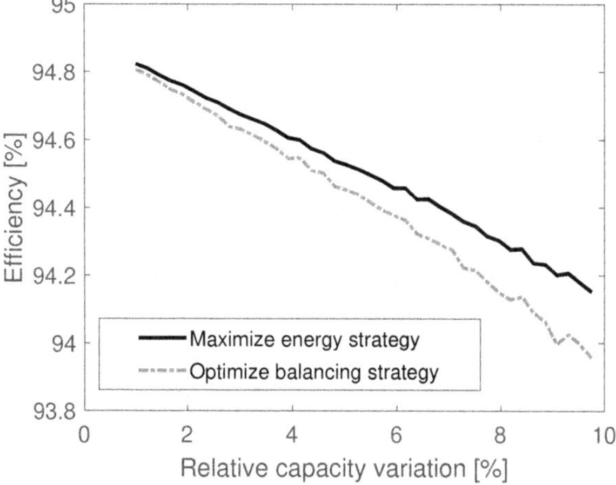

Figure 8.7: Comparison of the energy efficiency of the reconfigurable battery with the developed causal strategies. The used power profile represents the discharge phase of the FTP-75 driving cycle.

It is clear that the MES has slightly better results than OBS. This difference is because the OBS computes the number of active cells that solves the optimization problem (7.10a) regardless of the effect of the calculated number of active cells on energy efficiency.

8.2.3 Comparison of rest energy in reconfigurable and conventional batteries

Figure 8.7 shows the amount of energy that remains unusable at the end of discharge for three different systems: reconfigurable battery implementing the OBS, reconfigurable battery implementing *one-cell-off strategy*, and conventional battery implementing passive balancing.

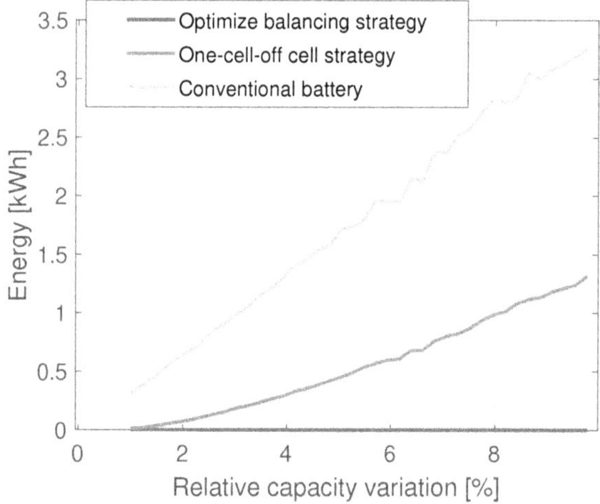

Figure 8.8: Comparison of the rest energy in the reconfigurable battery implementing different energy management strategies and conventional battery with fixed cell topology. The used power profile represents the discharge phase of the FTP-75 driving cycle.

The results show that the worse the capacity spread, the more energy remains in the conventional battery. For a relative capacity variation of about 8%, the remaining energy in the conventional battery is by 2.9kWh. The same effect can be observed by a reconfigurable battery implementing the state-of-the-art strategy *one-cell-off strategy*: the worse the capacity spread, the more energy remains at the end of discharge. However, when implementing the OBS, all cells are fully discharged so that no energy remains unusable in the battery.

8.2.4 Efficiency comparison with the one-cell-off strategy

Another case study relevant to assess the performance of the causal energy management strategies is the comparison with the state-of-the-art strategy, which is *one-cell-off strategy.* Figure 8.9 compares the energy efficiency of reconfigurable batteries implementing these strategies.

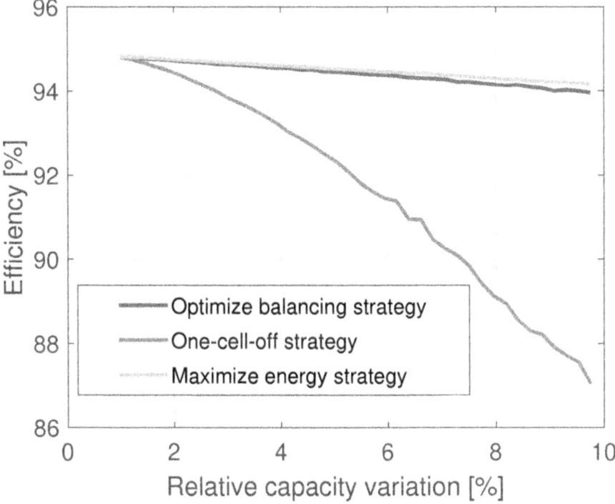

Figure 8.9: Comparison of the energy efficiency of the reconfigurable battery with different energy management strategies: EMS, OBS, and *one-cell-off strategy.* The used power profile represents the discharge phase of the FTP-75 driving cycle.

The causal models offer a significant improvement in the performance of the reconfigurable battery in comparison with the *one-cell-off strategy.*

8.3 Conclusion

The energy management strategies developed based on a static optimization model are well suited to be implemented in online applications since they do not require future information on the driving cycle. In addition, OBS is comparatively simple and can be easily implemented on battery management systems compared to the MES.

Chapter 9

Conclusion and outlook

The reconfigurable architecture is considered a promising solution against cell to cell variation inside multicell storage systems. Such architecture offers more system reliability and efficiency. The profit gained by the hardware architecture can be further enhanced when the control of the cells is optimized regarding the energy efficiency of the battery. Therefore, this work aims to develop energy management strategies using the optimization theory to maximize the energy efficiency of the battery.

First, the reconfigurable architecture tackled in this work is mathematically modeled. Namely, the impact of the number of active cells on the depth of discharge of each battery cell using mathematical formula is described. Besides, a mathematical model is provided that calculates the battery balancing state as a function of the number of active cells.

Using the optimization theory, the developed mathematical models are used to formulate three different energy management strategies. First, an optimal control problem in discrete-time form is introduced to compute the global optimum solution. This control problem is solved using an optimization algorithm developed based on the forward dynamic programming technique. The output of the algorithm is an optimal control policy over the given driving cycle. Using the equivalent circuit battery model and the calculated policy, simulation studies are conducted to assess the performance of the developed energy management strategy.

Results show that the optimal control strategy works very well regardless of the cell spread in the battery. In fact, for different cell-to-cell variation values, the energy efficiency of the reconfigurable battery is very close to that of a battery with uniform cells. This strategy requires that the complete driving cycle is known prior to the optimization process. Therefore, it is a noncausal strategy, and so it is not suitable for online applications. Nevertheless, it is used to benchmark the performance of the coming causal strategies.

In the next step, two causal energy management strategies are developed. They are a simplification of the noncausal strategy and do not require prediction of the future driving conditions. The first causal strategy is called *maximize energy strategy* (MES). It takes the current battery power requirement into consideration by determining the optimal control. This strategy maximizes the instantaneous benefit offered by reconfigurable architecture using the developed mathematical models.

The second causal strategy is called *optimize balancing strategy* (OBS). It only considers the current state of battery cells when calculating the optimal number of active cells. The corresponding static optimization problem is derived using the mathematical model that describes the impact of the number of active cells on each cell's depth of discharge. This strategy aims to find the number of active cells such that each cell in the reconfigurable battery reaches 100% depth of discharge and such that the amount of energy lost in the form of heat during battery operation is minimized. The OBS is computationally more efficient than the MES.

Simulation results demonstrate that both strategies exhibit good performance with respect to the energy efficiency enhancement and robustness against capacity variance. Namely, both allowed the reconfigurable battery to be operated almost along the energy efficiency values achieved by the noncausal strategy. Whereby the MES provides slightly better results which makes it the favorite strategy.

The optimization-based strategies proposed in this work allow a significant enhancement in energy efficiency over that of a conventional battery, especially by increased capacity spread. In addition, a comparison with the performance of a rule-based strategy shows that the benefit regarding electrical energy gained due to the reconfigurable architecture is much further enhanced when an optimization-based strategy is used.

Last but not least, reconfigurable architecture could have a positive effect on the aging of Lithium-ion batteries. Namely, the proposed strategies activate cells with more energy content more often. As a result, these cells have more charge throughput than those with lower capacity. So, cells age differently in such a way that weak cells show a slower capacity fade over the lifetime. Consequently, due to the novel battery architecture, cells are operated appropriately so that the inhomogeneity of battery modules will be reduced and the battery lifetime will be enhanced.

Nomenclature and Abbreviations

ξ	State of charge of a cell	[%]
$\xi_{e,j}$	State of charge of the cell j at the end of discharge	[%]
E_{ext}	The amount of energy left in the battery during the operation	[J]
E_j	The total energy content of the cell j	[J]
$\boldsymbol{u}$	Control variable	[·]
$\boldsymbol{x}$	State variable	[·]
$\boldsymbol{\lambda}$	Lagrange multiplier	[·]
$\boldsymbol{\mu}$	Lagrange multiplier	[·]
$\boldsymbol{i}_{\text{cell}}$	Vector that defines the current values of battery cells	[A]
$\boldsymbol{i}_j$	Current of the cell j	[A]
$\boldsymbol{m}$	Vector that indicates which of the battery cells are fully discharged at time t_{e}	[·]
$\boldsymbol{s}$	Vector that defines which cells are connected	[·]
ξ_0	The initial state of charge	[%]
B	Set of the battery cell capacities	[·]
C	The effective usable cell capacity	[As]

E_{bcg}	Equalization energy	[J]
E_l	Energy loss inside the battery	[J]
$f(\cdot)$	System dynamic function	[$\cdot$]
$f_E(\cdot)$	Extracted energy function	[$\cdot$]
$H(\cdot)$	Hamilton function	[$\cdot$]
i_{bat}	Battery current	[A]
$J(\cdot)$	Cost functional	[$\cdot$]
k	Stage	[$\cdot$]
L	Lagrangian function	[$\cdot$]
$l(\cdot)$	Running cost function	[$\cdot$]
M	The number of cells that have 100% depth of discharge	[$\cdot$]
N	Number of battery cells	[$\cdot$]
n_{on}	The number of active cells	[$\cdot$]
$P_{l,j}$	Power loss of the cell j	[W]
$Q_{\text{th,bat}}(t)$	Battery charge throughput	[As]
$Q_{\text{th},j}(t)$	Charge throughput of the cell j	[As]
$Q_{0,j}$	The charge of the cell j at the beginning of discharge	[As]
R_0	Cell ohmic resistance	[Ω]
Ri	Cell internal resistance	[Ω]
T	Final stage	[$\cdot$]

t_e	The end of discharge time	[s]
$V(\cdot)$	Final cost function	$[\cdot]$
V_{oc}	Open circuit voltage	[V]
χ	Set of number of active cells	$[\cdot]$
KKT	Karush–Kuhn–Tucker conditions	$[\cdot]$

List of Tables

List of Figures

Bibliography

[1] S. Lehner, T. Baumhöfer, and D. U. Sauer, "Disparity in initial and lifetime parameters of lithium-ion cells", IET Electrical Systems in Transportation **6**, 34–40 (2016).

[2] L. Lu, X. Han, J. Li, J. Hua, and M. Ouyang, "A review on the key issues for lithium-ion battery management in electric vehicles", Journal of Power Sources **226**, 272–288 (2013).

[3] J. Gallardo-Lozano, E. Romero-Cadaval, M. I. Milanes-Montero, and M. A. Guerrero-Martinez, "Battery equalization active methods", Journal of Power Sources **246**, 934–949 (2014).

[4] A. Manenti, A. Abba, A. Merati, S. M. Savaresi, and A. Geraci, "A new BMS architecture based on cell redundancy", IEEE Transactions on Industrial Electronics **58**, 4314–4322 (2011).

[5] A. Sciarretta and L. Guzzella, "Control of hybrid electric vehicles", IEEE Control Systems Magazine **27**, 60–70 (2007).

[6] L. Guzzella and A. Sciarretta, *Vehicle propulsion systems: introduction to modeling and optimization* (Springer Science Business Media, 2005).

[7] C. Zhu and B. Yang, "Energy management in parallel hybrid vehicle using fuzzy control", in IEEE International Conference on Information Science and Technology (2012).

[8] W. Liu, "Energy management strategies for hybrid electric vehicles", in *Hybrid Electric Vehicle System Modeling and Control, Second Edition* (John Wiley Sons Ltd, 2017), pp. 243–287.

[9] R. Biasini, S. Onori, and G. und Rizzoni, "A near-optimal rule-based energy management strategy for medium duty hybrid truck", International Journal of Powertrain **2**, 232–261 (2013).

[10] D. Ambühl, L. Guzzella, and A. Sciarretta, "Explicit optimal control policy and its practical application for hybrid electric powertrains", Control Engineering Practice **18**, 1429–1439 (2010).

[11] D. Moosbauer, "Elektrochemische charakterisierung von elektrolyten und elektroden für Lithium-ionen-batterien", PhD thesis (Regensburg Universtity, 2010).

[12] G. Nazri and G. Pistoia, *Lithium batteries science and technoloy* (Springer US, 2009).

[13] M. Osiak, H. Geaney, E. Armstrong, and C. O'Dwyer, "Structuring materials for Lithium-ion batteries: advancements in nanomaterial structure, composition, and defined assembly on cell performance", Journal of Materials Chemistry A **2**, 9433–9460 (2014).

[14] C. Liu, Z. Neale, and G. Gao, "Understanding electrochemical potentials of cathode materials in rechargeable batteries", Materials Today **2**, 109–123 (2016).

[15] R. Korthauer, *Lithium-ion batteries: basics and applications* (Springer, Berlin, Heidelberg, 2018).

[16] R. Korthauer, *Handbuch Lithium-ionen Batterien* (Springer, Heidelberg, 2013).

[17] R. Rao, S. Vrudhula, and D. N. Rakhmatov, "Battery modeling for energy-aware system design", Computer **36**, 77–87 (2003).

[18] F. Saidani, F. X. Hutter, R.-G. Scurtu, W. Braunwarth, and J. N. Burghartz, "Lithium-ion battery models: a comparative study and a model-based powerline communication", Advances in Radio Science **15**, 83–91 (2017).

[19] M. Ecker, S. Kaebtiz, I. Laresgoiti, and D. U. Sauer, "Parameterization of a physico-chemical model of a Lithium-ion battery", Journal of The Electrochemical Society **169**, 1849–1857 (2015).

[20] H. He, R. Xiong, and J. Fan, "Evaluation of Lithium-ion battery equivalent circuit models for state of charge estimation by an experimental approach", Energies **4**, 582–598 (2011).

[21] L. W. Juang, P. J. Kollmeyer, T. M. Jahns, and R. D. Lorenz, "Improved modeling of Lithium-based batteries using temperature-dependent resistance and overpotential", in International Conference on Electric Information and Control Engineering (ICEICE) (2014).

[22] I. Dincer and M. A. Rosen, *Exergy: energy, environment and sustainable development* (Elsevier Ltd, 2007).

[23] M. Chen and G. A. Rincon-Mora, "Accurate electrical battery model capable of predicting runtime and I-V performance", IEEE Transactions on Energy Conversion **21**, 504–511 (2006).

[24] T. Kim, W. Qiao, and L. Qu, "Series-connected reconfigurable multicell battery: a novel design towards smart batteries", in IEEE Energy Conversion Congress and Exposition (ECCE) (2010).

[25] P. Zhang, C. Du, F. Yan, and J. Kang, "Influence of practical complications on energy efficiency of the vehicle's Lithium-ion batteries", in International Conference on Electric Information and Control Engineering (ICEICE) (2011).

[26] A. Kampker, H. H. Heimes, M. Ordung, C. Lienemann, A. Hollah, and N. Sarovic, "Evaluation of a remanufacturing for Lithium ion batteries from electric cars", International Journal of Mechanical and Mechatronics Engineering **12**, 1929–1935 (2016).

[27] T. Bandhauer, S. Garimella, and T. Fuller, "Critical reviews in electrochemical and solid-state science and technology", Journal of The Electrochemical Society.

[28] V. Srinivasan and C. Y Wang, "Analysis of electrochemical and thermal behavior of li-ion cells", Journal of The Electrochemical Society **150**, 98–106 (2003).

[29] J. Cao, N. Schofield, and A. Emadi, "Battery balancing methods: a comprehensive review", in IEEE Vehicle Power and Propulsion Conference (VPPC) (2008).

[30] M. Caspar, T. Eiler, and S. Hohmann, "Comparison of Active Battery Balancing Systems: A Model-Based Quantitative Analysis", IEEE Transactions on Vehicular Technology **67**, 920 –934 (2018).

[31] M. Daowd, M. Antoine, N. Omar, P. van den Bossche, and J. van Mierlo, "Single switched capacitor battery balancing system enhancements", Energies **6**, 2149–2174 (2013).

[32] Z. Gao, C. S. Chin, J. Jia, W. Toh, and J. Chiew, "State-of-charge estimation and active cell pack balancing design of Lithium battery power system for smart electric vehicle", Journal of Advanced Transportation **2017**, 1–14 (2017).

[33] M. Daowd, M. Antoine, N. Omar, P. van den Bossche, and J. van Mierlo, "Passive and active battery balancing comparison based on matlab simulation", in IEEE Vehicle Power and Propulsion Conference (VPPC) (2011).

[34] M. Caspar and S. Hohmann, "Optimal cell balancing with model-based cascade control by duty cycle adaption", in The International Federation of Automatic Control (2014).

[35] M. samadi, Maurice, and M Saif, "Nonlinear model predictive control for cell balancing in Li-ion battery packs", in American Control Conference (ACC) (2014).

[36] M. Einhorn, W. Guertlschmid, T. Blochberger, R. Kumpusch, R. Permann, F. V. Conte, C. Kral, and J. Fleig, "A current equalization method for serially connected battery cells using a single power converter for each cell", IEEE Transactions on Vehicular Technology **60**, 4227–4237 (2011).

[37] W. Huang and J. A. A. Qahouq, "Distributed battery energy storage system architecture with energy sharing control for charge balancing", in Applied Power Electronics Conference and Exposition (APEC) (2014).

[38] W. Huang and J. A. A. Qahouq, "Energy sharing control scheme for state-of-charge balancing of distributed battery energy storage system", IEEE Transactions on Industrial Electronics **62**, 2764–2776 (2015).

[39] N. Bouchhima, M. Gossen, S. Schulte, and K. Birke, "Fundamental aspects of reconfigurable batteries: efficiency enhancement and lifetime extension", in *Modern Battery Engineering* (2019).

[40] T. Zimmermann, P. Keil, M. Hofmann, M. F. Horsche, S. Pichlmaier, and A. Jossen, "Review of system topologies for hybrid electrical energy storage systems", Journal of Energy Storage **8**, 78–90 (2016).

[41] H. Kim and K. G. Shin, "On dynamic reconfiguration of a large-scale battery system", in 15th IEEE Real-Time and Embedded Technology and Applications Symposium (2009).

[42] S. K. Mandal, P. S. Bhojwani, S. P. Mohanty, and R. N. Mahapatra, "Intellbatt: towards smarter battery design", in 45th ACM/IEEE Design Automation Conference (2008).

[43] T. Kim, W. Qiao, and L. Qu, "A multicell battery system design for electric and plug-in hybrid electric vehicles", in IEEE International Electric Vehicle Conference (2012).

[44] T. Kim, W. Qiao, and L. Qu, "Power electronics-enabled self-x multicell batteries: a design toward smart batteries", IEEE Transactions on Power Electronics **27**, 4723–4733 (2012).

[45] L. He, L. Gu, L. Kong, Y. Gu, C. Liu, and T. He, "Exploring adaptive reconfiguration to optimize energy efficiency in large-scale battery systems", in IEEE 34th Real-Time Systems Symposium (2013).

[46] V. Muenzel, J. de Hoog, M. Brazil, D. A. Thomas, and I. Mareels, "Battery management using secondary loads: a novel integrated approach", in 19th World CongressThe International Federation of Automatic Control (2014).

[47] F. Baronti, G. Fantechi, R. Roncella, and R. Saletti, "Design of a module switch for battery pack reconfiguration in high-power applications", in IEEE International Symposium on Industrial Electronics (2012).

[48] A. Klieber, "Untersuchung von Betriebsstrategien für Einzelzelltopologie Batterien", Master thesis (University of Stuttgart, 2012).

[49] T. Kim, W. Qiao, and L. Qu, "A series-connected self-reconfigurable multicell battery capable of safe and effective charging/discharging and balancing operations", in IEEE Applied Power Electronics Conference and Exposition (2012).

[50] T. Kim, W. Qiao, and L. Qu, "Series-connected self-reconfigurable multicell battery", in IEEE Applied Power Electronics Conference and Exposition (APEC) (2011).

[51] T. Uesugi, "Power devices for automotive applications-review of technologies for low power dissipation and high ruggedness", RD Review of Toyta CRDL **35**, 1–7 (2000).

[52] D. Görke, *Untersuchungen zur kraftstoffoptimalen Betriebsweise von Parallelhybridfahrzeugen und darauf basierende Auslegung regelbasierter Betriebsstrategien* (Springer, Viewweg, 2015).

[53] S Wahsh, H. G. Hamed, M. N. F. Nashed, and T Dakrory, "Fuzzy logic based control strategy for parallel hybrid electric vehicle", in 11th International Middle East Power Systems Conference (2006).

[54] I. Arsie, M. Graziosi, C. Pianese, G. Rizzo, and M. Sorrentino, "Optimization of supervisory control strategy for parallel hybrid vehicle with provisional load estimate", in Advanced Vehicle Control AVEC (2004).

[55] A. Sciarretta, L. Guzzella, and M. Back, "A real-time optimal control strategy for parallel hybrid vehicles with on-board estimation of the control parameters", in IFAC Symposium on Advances in Automotive Control (2004), pp. 489–494.

[56] A. Fill, S. Koch, A. Pott, and K. P. Birke, "Current distribution of parallel-connected cells in dependence of cell resistance, capacity and number of parallel cells", Journal of power sources **407**, 147–152 (2018).

[57] K. Graichen, *Methoden der optimierung und optimalen steuerung*, Lecture notes at the University of Ulm, 2015.

[58] D. E. Marthaler, "An overview of mathematical methods for numerical optimization", in *Numerical methods for metamaterial design* (Springer, Netherlands, 2013), pp. 31–53.

[59] D. Liberzon, *Calculus of variations and optimal control theory* (Princeton University Press, 2011).

[60] H. Schättler and U. Ledzewicz, *Geometric optimal control theory, methods and examples* (Springer-Verlag New York, 2012).

[61] M. Papageorgiou, M. Leibold, and M. Buss, *Optimierung Statische, dynamische, stochastische Verfahren für die Anwendung* (Springer Vieweg, 2015).

[62] S. Sager, "Numerical methods for mixed–integer optimal control problems", PhD thesis (Universit at Heidelberg, 2006).

[63] A. Banitalebi, R. Ahmad, and M. I. Abd Aziz, "A direct method for optimal control problem", International Journal of Pure and Applied Mathematics **81**, 33–47 (2012).

[64] P. A. Jensen and J. F. Bard, *Operations research models and methods* (John Wikey Sons. Inc., Hoboken, NJ, 2003).

[65] A. A. Sitinjak, E. Pasaribu, J. E. Simarmata, T. Putra, and H. Mawengkang, "The analysis of forward and backward dynamic programming for multistage graph", in 4th International Conference on Operational Research (2017), pp. 1–7.

[66] D. S. Bayard, "Reduced complexity dynamic programming based on policy iteration", Journal of Mathematical analysis and applications **170**, 75–103 (1991).

[67] L. Cheng Yu Huei, "Control strategy optimization for parallel hybrid electric vehicles using a memetic algorithm", Energies **10**, 1–21 (2003).

[68] M. Montazeri-Gh, A. Poursamad, and B. Ghalichi, "Application of genetic algorithm for optimization of control strategy in parallel hybrid electric vehicles", Journal of the Franklin Institute **343**, 420–435 (2006).

[69] J Dantas De Melo, J. Calvet, and J. Garcia, "Vectorization and multitasking of dynamic programming in control: experiments on a CRAY-2", Parallel Computing, 261–269 (1990).

[70] T. Kim, W. Qiao, and L. Qu, "Real-time state of charge and electrical impedance estimation for lithium-ion batteries based on a hybrid battery model", in 28th IEEE Applied Power Electronics Conference and Exposition, edited by editor (2013).

[71] S. Paul, "Analyse der Ausfallwahrscheinlichkeiten von Lithium-ionen Energiespeichern in elektrifizierten Fahrzeugen", PhD thesis (ULM universitity, 2015).

[72] D Andre, M Meiler, K. Steiner, C. Wimmer, T Soczka-Guth, and D. U. Sauer, "Characterization of high-power Lithium-ion batteries by electrochemical impedance spectroscopy. ii: modelling", Journal of power sources **196**, 5349–5356 (2011).

[73] T. Baumhöfer, M. Brühl, S. Rothgang, and D. U. Sauer, "Production caused variation in capacity aging trend and correlation", Journal of Power Sources **247**, 332–338 (2014).

[74] S. Paul, C. Diegelmann, H. Kabza, and W. Tillmetz, "Analysis of ageing inhomogeneities in Lithium-ion battery systems", Journal of Power Sources **239**, 642–650 (2013).

[75] S. F. Schuster, M. J. Brand, P. Berg, and M. Gleissenberger, "Lithium-ion cell-to-cell variation during battery electric vehicle operation", Journal of Power Sources **297**, 242–251 (2015).

[76] L. Zhou, Y. Zheng, M. Ouyang, and L. Lu, "A study on parameter variation effects on battery packs for electric vehicles", Journal of Power Sources **364**, 242–252 (2017).

[77] S. Rothgang, T. Baumhöfer, and D. U. Sauer, "Necessity and methods to improve battery lifetime on system level", in Kintex (2015).

[78] W. GIS, Żółtowski Andrzej, and A. Bocheńka, "Testing of the electric vehicle in driving cycles", Journal of KONES Powertrain and Transport **19**, 207–221 (2012).

[79] W. Moćko and W. Gis, "Development and validation of model of the electric car energy consumption", PRZEGLAD ELEKTROTECHNICZNY **89**, 141–143 (2013).

[80] G. Paganelli, S. Delprat, T. M. Guerra, J. Rimaux, and J. J. Santin, "Equivalent consumption minimization strategy for parallel hybrid powertrains", in IEEE 55th Vehicular Technology Conference (2002).

[81] J. W. Chinneck, *Practical optimization: a gentle introduction*, Lecture notes at the University of Carleton, 2016.

Contributions

The following scientific publications were produced within the framework of this thesis:

1. Nejmeddine Bouchhima, Matthias Gossen and Kai Peter Birke; Fundamental Aspects of Reconfigurable Batteries: Efficiency Enhancement and Lifetime Extension; Modern Battery Engineering (2019) 101-119.
2. Nejmeddine Bouchhima, Matthias Gossen, Sascha Schulte and Kai Peter Birke; Lifetime of self-reconfigurable batteries compared with conventional batteries; Journal of Energy Storage 15 (2018) 400-407.
3. Nejmeddine Bouchhima, Marc Schnierle, Sascha Schulte and Kai Peter Birke; Optimal energy management strategy for self-reconfigurable batteries; Journal of Energy 322 (2017) 129–137.
4. Nejmeddine Bouchhima, Marc Schnierle, Sascha Schulte and Kai Peter Birke; Active model-based balancing strategy for self-reconfigurable batteries; Journal of Power Sources 322 (2016) 129–137.

Curriculum Vitae

Personal Data

Name	Nejmeddine Bouchhima
Date of birth	12.01.1987
Place of birth	Sfax, Tunisia

Work Experience

09/2018 - present	System Engineer Advanced Driver Assistance Systems, Daimler AG
03/2017 - 08/2018	System Engineer Lithium-ion Batteries for electric vehicles, Daimler AG

Education

10/2006 - 09/2008	Studies in Mechanical Engineering at Karlsruhe Institute of Technology, intermediate diploma
10/2008 - 09/2012	Studies in Mechatronic at Karlsruhe Institute of technology, Diploma in September 2012
10/2010 - 03/2011	Internship at Robert Bosch, Reutlingen
02/2013 - 02/2017	Doctoral student at Institute for Photovoltaics, University of Stuttgart

Erklärung

Ich versichere hiermit, dass ich die vorliegende Arbeit selbstständig verfasst und keine anderen als die angegebenen Hilfsmittel verwendet habe.

Stuttgart, 10. Januar 2020

Nejmeddine Bouchhima

Acknowledgments

First of all, I would like to thank my supervisor, Prof. Dr. Kai Peter Birke, for the continuous support throughout my doctoral study, for the confidence he expressed in me, and for the valuable technical discussions. His guidance helped me in all the time of research and writing of this thesis. Further, I am grateful to Prof. Dr.-Ing. Julia Kowal who accepted to be my co-examiner, for the encouragement and the insightful comments.

Further, I wish to thank all my colleagues for the great working environment and the in-depth technical discussions. Special thanks go to Dr.-Ing. Sascha Schulte for the fruitful technical discussions and for proofreading my publications. I also want to thank Matthias Gossen and Marc Schnierle for the excellent collaboration, resulting in joint publications.

Finally, I would like to express my overwhelming gratitude to my family. Its support and encouragement were important to me to realize this dissertation.

www.ingramcontent.com/pod-product-compliance
Ingram Content Group UK Ltd.
Pitfield, Milton Keynes, MK11 3LW, UK
UKHW022000190726
13853UKWH00004B/1650